**Studies in
Canadian
Geograph**

**Etudes su
la géographie**

Date Due

APR 20

917.15 Macpherson, Alan Gibson, 1927-   ed.
Macp   The Atlantic Provinces; Les Provinces de
        l'Atlantique. Ed. by Alan Macpherson.
        Toronto, University of Toronto Press, 1972.
          xi, 182 p. illus., maps. (Studies in
        Canadian geography)
          Published for the 22nd International
        Geographical Congress, Montreal, 1972.
          Bibliography: p. 177-182.

                        (Continued on next card)
                              G-8604
                              G-9444

The Atlantic Provinces

Les Provinces de l'Atlantique

Edited by /Sous la direction de Alan Macpherson

published for the 22nd International Geographical Congress
publié à l'occasion du 22e Congrès international de géographie
Montréal 1972

University of Toronto Press

© University of Toronto Press 1972
Toronto and Buffalo

ISBN 0-8020-1916-1 (Cloth)
ISBN 0-8020-6158-3 (Paper)
Microfiche ISBN 0-8020-0255-2

Printed in Canada

Contents

Foreword

The publication of the series, 'Studies in Canadian Geography,' by the organizers of the 22nd International Geographical Congress, introduces to the international community of geographers a new perspective of the regional entities which form this vast country. These studies should contribute to a better understanding among scholars, students, and the people of Canada of the geography of their land.

Geographical works embracing the whole of Canada, few in number until recently, have become more numerous during the last few years. This series is original in its purpose of re-evaluating the regional geography of Canada. In the hope of discovering the dynamic trends and the processes responsible for them, the editors and authors of these volumes have sought to interpret the main characteristics and unique attributes of the various regions, rather than follow a strictly inventorial approach.

It is a pleasant duty for me to thank all who have contributed to the preparation of the volume on the Atlantic Provinces. A special thanks is due to: Mr R.I.K. Davidson of the University of Toronto Press; Mr Geoffrey Lester who guided the Cartography Laboratory of the Department of Geography, University of Alberta in preparing all the illustrations; the Canadian Association of Geographers for its financial support; and the Executive of the Organizing Committee of the 22nd International Geographical Congress. Finally I wish to thank Dr Alan G. Macpherson, professor of geography at the Memorial University of Newfoundland, for having accepted the editorship of this volume.

LOUIS TROTIER
Chairman
Publications Committee

Avant-propos

Par la publication de cette série d''Etudes sur la géographie du Canada,' les organisateurs du 22e Congrès international de géographie ont voulu profiter de l'occasion qui leur était donnée de présenter à la communauté internationale des géographes une perspective nouvelle des grands ensembles régionaux qui composent cet immense pays. Ils espèrent que ces études contribueront aussi à mieux faire comprendre la géographie de leur pays aux Canadiens eux-mêmes, scientifiques, étudiants ou autres.

Les travaux d'ensemble sur la géographie du Canada, peu nombreux jusqu'à récemment, se sont multipliés au cours des dernières années. L'originalité de cette série provient surtout d'un effort de renouvellement de la géographie régionale du Canada. Les rédacteurs et les auteurs de ces ouvrages ont cherché moins à inventorier leur région qu'à en interpréter les traits majeurs et les plus originaux, dans l'espoir de découvrir les tendances de leur évolution.

C'est pour moi un agréable devoir de remercier et de féliciter tous ceux qui ont contribué d'une manière ou d'une autre à la réalisation de cet ouvrage sur les provinces de l'Atlantique. Il convient de mentionner les membres du Comité d'organisation du 22e Congrès international de géographie; M. R.I.K. Davidson, des Presses de l'Université de Toronto; l'Association canadienne des géographes; le département de géographie de l'Université de l'Alberta, à Edmonton, dont le Laboratoire de cartographie a préparé toutes les ilustrations de cet ouvrage sous la direction habile et dévouée de M. Geoffrey Lester. Je remercie enfin M. Alan G. Macpherson, professeur de géographie à l'Université Memorial de Terre-Neuve, d'avoir accepté d'assumer la direction de cet ouvrage.

LOUIS TROTIER
Président du
Comité des publications

Preface

The Atlantic Provinces of Canada – Newfoundland and the three Maritime Provinces of Nova Scotia, New Brunswick, and Prince Edward Island – hold a special place in the affection of their sister provinces across Canada. The mainland provinces of Nova Scotia and New Brunswick were part of the original confederation with Quebec and Ontario in 1867; Prince Edward Island joined as the sixth province in 1873 (after British Columbia); and Newfoundland, after hesitating much longer, completed confederation in 1949 and allowed the heraldry on the outer wall of Parliament in Ottawa to be finished.

People of the Atlantic Provinces are found in every province and territory and have been in recent times, in fact, the most numerous group of 'immigrants' moving into the national territory west of the Madawaska and the Labrador Trough, except for the English. In 1961 over 300,000 of them were recorded in Canada outside their home region, which at that time had a population of two million and had also been contributing heavily to the population of the United States. American-born residents in Canada in that year numbered but 284,000. The Atlantic region's contribution to the present state of the rest of Canada has therefore been considerable in social and economic as well as demographic terms.

In its varied folk traditions the region adds to the national scene a cultural dimension which has all but vanished elsewhere in rural Canada, at least among the English, Irish, and Scots Canadians. The strength of the rural tradition is matched by a countryside – from the Madawaska to the Southern Shore of the Avalon southward of St John's and from the French Shore of southern Nova Scotia to northern Labrador – which in all its variety is one of the loveliest in Canada, and even yet one of the least spoiled. It is probably the comfort of these two aspects of life, rather than any purely economic advantage, which has promoted a strong and persistent return of out-migrant sons and daughters to the Atlantic Provinces. This has been identified for Prince Edward Island in particular for the 1956–61 period and in the 30–44-year-group (Lycan), but the net figures conceal the full extent of this in terms of numbers and the other provinces involved. Easy access to common resources is part of the tradition, and the distinction between recreational activity and exploitation for subsistence is a blurred one.

In purely economic terms the region is reckoned to be one of disparity as compared with the rest of Canada. Educational levels in the population, though rising, are still significantly below those in other parts of the country. Comparative measures developed by the Economic Council of Canada indicate that between 1961 and 1964 the average income per person in the Atlantic Provinces was only 66 per cent, the average income per employed person only 80 per cent, of the Canadian average. (The average income per Newfoundlander in 1969 was 56 per cent of the national average.) The first figure can be related to an age-structure more heavily weighted in the younger non-earning age-groups; the second figure, in part at least, reflects differences in educational levels, with consequent differences in type of employment and general level of development. Between 1961 and 1964 only 47.4 per cent of the population participated in the labour force, as compared with 54 per cent in the country as a whole: again a reflection of a different demographic structure and general level of development. For men, participation ran at 71.8, for women 23.6; cf 78.9 and 29.5 per cent. In Newfoundland in 1970 the participation rate was 42.9 per cent (men: 61.4; women: 24.3 per cent). Unemployment rates are always expected to be the highest of any region in Canada, no matter the level of national unemployment, and seasonal variations in employment rates are generally twice the national fluctuation (1961–64: 8.6 per cent to 4.4 per cent nationally). The image created by the economists, nevertheless, while truer for the urban sector, is only a dangerous half-truth for the region as a whole and a peril to national planning agencies and politicians.

In terms of scientific interest the region has assumed a prominent position in a number of fields during the last decade. Dr Anne Stine's excavations at L'Anse aux Meadows (L'Anse aux Meduses) on the northern tip of Newfoundland in the early 1960s have finally provided archaeological evidence for Norse attempts to colonize a North American shore. The somewhat variable carbon-14 dates (A.D. 1090 ± 90; A.D. 890 ± 70) are just right once they have been corrected from the Californian bristlecone pine calibration. The site is in process of becoming part of a National Historic Park. Cartographic interest was added in 1965 with the announcement of the discovery and authentification of the Vinland Map, a suggested date of c. 1440, and arguments for assuming the existence of much earlier prototypes in the North European tradition of map-making.

More exciting and archaeologically significant were Dr James Tuck's excavations of a Maritime Archaic Indian cemetery (2340–1280 B.C.) at Port aux Choix on Newfoundland's Northern Peninsula in 1967. The tool kits recovered, as well as the skeletal remains, represent a haul richer than any found in Maine or the Maritimes. Extension of his work to Northern Labrador is continuing to establish the importance of this cultural complex within North American prehistory.

A third field which has recently brought the region to the attention of scholars lies in military history and archaeology both above and below the tide mark, of which perhaps the investigations which have gone into the reconstruction of the great French naval base at Louisbourg on Cape Breton Island are the most significant.

And finally it should be noted that the Atlantic Provinces, and Newfoundland in particular, lie central in the field evidence for the new theories of plate tectonics and sea-floor spreading which are currently revolutionizing the geological and geophysical sciences.

The Atlantic Provinces can be thought of as a region where three zones of primary exploitation meet: the Northern North Atlantic, a zone of coastal settlement marginal to waters heavily fished by international fleets; the continental Sub-Arctic, a zone of Boreal Forest exploitation dominated by international pulp and paper companies and dependent upon highly competitive world markets; and Appalachia, a zone of mineral exploitation involving metallic and non-metallic ores and fossil fuels highly dependent upon international financing and marketing. Each of the economies associated with these zones has brought into the region problems of environmental destruction and control, economic stagnation and development, and social decline and rehabilitation. Insofar as industry is fundamentally and ultimately irresponsible in its relationship with the people of any region where exploitation of resources is underway or intended, prospects for the Atlantic Provinces must be based ultimately upon political decisions. And it is in the area of political decisions that the region's basic weakness lies.

The international nature of the problems faced by the Atlantic Provinces – an old theme in Canadian Maritime history – requires that the federal government of Canada seek solutions on their behalf. Room to manoeuvre within their constitutional jurisdictions over resources and people is too limited to effect much if there is no federal will or initiative in such fields as international finance, management of marine resources, and pollution of the sea, or in such national areas as environmental control, processing and marketing policy, and population planning. Weakness in obtaining federal initiative on their behalf springs from the political fragmentation of the Atlantic Provinces and the almost total lack of a regional consciousness. Prospects for the region, therefore, ultimately depend upon its overcoming this condition, which, as this volume has demonstrated, springs from a fundamental fragmentation in its geography. No attempt has been made to present a comprehensive geography of the Atlantic Provinces, but the four essays endeavour to present this part of Canada as essentially a region of geographical fragmentation.

Memorial University of Newfoundland   A.G.M.
14 June 1972

'... by those unchanging laws of geological structure and geographical position which the Creator himself has established, this region must always, notwithstanding any artificial arrangements that man may make, remain distinct from Canada on the one hand, and New England on the other ... The resources of the Acadian provinces must necessarily render them more wealthy and populous than any area of the same extent on the Atlantic coast, from the Bay of Fundy to the Gulf of Mexico, or in the St. Lawrence valley, from the sea to the head of the great lakes. Their marine and mineral resources constitute them the Great Britain of Eastern America; and though merely agricultural capabilities may give some inland and southern regions a temporary advantage, Acadia will in the end assert its natural pre-eminence.' J.W. Dawson, 1868

1 The Physical Geography of the Atlantic Provinces

IAN BROOKES

Perhaps it is because it is too well known that the physical geography of the Canadian Atlantic Provinces[1] remains poorly understood. Accounts of inhospitable highlands, a cold and wet climate, acid soils, mixed forests, etc. have adequately served to frame commonplace conceptions of the human geography of the region, but have resulted in a dearth of satisfying treatments. The rare exceptions have been concerned only with parts of the region and have intended only to preface discussions of its human geography (Clark 1968; Erskine 1968).

The lack of a comprehensive treatment cannot be fully explained by a paucity of background material. Indeed, the present writer has been confronted with an embarrassment of riches. While many different approaches could have been adopted, the theme of fragmentation followed in this volume helps to circumscribe the task. Although even the most cursory glance at the basic elements of physical geography in the Atlantic region will suggest fragmentation as a fruitful frame of reference, use of the word does beg a question about the real nature of interrelations between a fragmented physical-biotic environment and the people who inhabit it.

This chapter attempts to outline the dimensions, character, and genesis of elements of the physical-biotic environment of man in the Atlantic Provinces. It is avowedly not analytical, being more of a 'state of the art' review.

THE DIMENSIONS OF THE REGION

The Atlantic Provinces occupy an area of 95,322 square miles (275,480 km²) – 2.5 per cent of the Canadian land area, about one-fifth of Ontario, and about one-third of Alberta, Saskatchewan, or Manitoba. This small

1 In this chapter the term 'Atlantic Provinces' refers to New Brunswick, Prince Edward Island, Nova Scotia, and the island of Newfoundland. The Labrador section of Newfoundland is excluded, together with the Gaspé and Laurentian sections

area is, however, spread over great distances. The 800-mile (*ca.* 1300 km) distance west-to-east from northwestern New Brunswick to St. John's, Newfoundland, is comparable to the air distance between Toronto and Halifax, Calgary and Winnipeg, or Vancouver and Regina. Further, the eight-degree difference in latitude between Cape Sable, Nova Scotia, and Cape Bauld, Newfoundland, is the same as that between Winnipeg and Churchill, Manitoba.

This area-distance paradox arises out of the relatively large area of the island of Newfoundland, which, at 43,359 square miles (125,307 km²), is twice the size of Nova Scotia, and the even larger area of the Gulf of St Lawrence – roughly 57,000 square miles (164,730 km²). In addition to the land area, 220,000 square miles (635,800 km²) of continental shelf must be included within the Atlantic region for present purposes, since a large proportion of the population has traditionally gained a livelihood from its waters (Figure 1.1). Taken as a whole, the combined area of land and shelf seas in the region, 315,000 square miles, is still only three-quarters that of Ontario.

It is from a small-scale topographic map that the impression of fragmentation is most easily gained (Figure 1.1). Each province, except lowland Prince Edward Island, experiences abrupt transitions between upland blocks and lowland basins that are, in the main, structurally controlled. The upland aspect derives more from local relief than altitude: the highest point in the region, Mt Carleton, in northern New Brunswick, 2690 feet (820 m), affords spectacular views of surrounding plateaus and valleys, but is, nevertheless, some 800 feet (240 m) nearer sea level than the city of Calgary.

The uplands are disposed in two massive bulks, in northern New Brunswick and over most of Newfoundland, and in two more linear belts, one across southern New Brunswick and northern Nova Scotia, the other over peninsular Nova Scotia and Cape Breton Island. They have in common a plateau aspect, with smoothly undulating skylines, whether they stand at 500–800 feet (150–240 m) as in peninsular Nova Scotia, 800–1200 feet (240–360 m) in the central zone, or 2000–2500 feet (600–750 m) in northern New Brunswick and western Newfoundland. Plateau surfaces are broken by steep-sided valleys and are bounded by steep slopes developed along geological boundaries. It is difficult to resist a determinist evaluation of these uplands as effective barriers between low-

of Quebec. The term 'Maritime Provinces' is used to refer to New Brunswick, Prince Edward Island, and Nova Scotia. The term 'Atlantic Region' is used here synonymously with 'Atlantic Provinces,' and, where offshore areas are being discussed, they are included in the meaning of the term.

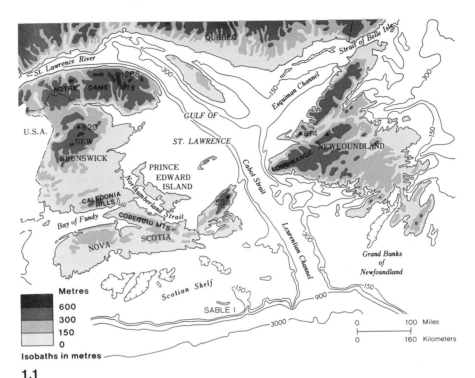

1.1

Bathy-Orography of the Atlantic Provinces

(Source: Atlas of Canada, 1957)

land-centred population clusters. There are few easy routes across them; many routes seem to hesitate before leading into them and return to the lowland corridors rather than climb across to meet their opposite numbers. Just as many are one-way routes to a 'head of gravel.'

The isolated valleys dissecting these uplands have acquired an importance to the dispersed farming population out of proportion to their potential. Bounded by steep sides in bedrock or lightly mantled with bouldery glacial till, they have a level floor, generally one half to one mile (0.8–1.6 km) wide, over which are spread sandy and loamy glacial materials that have served well in support of small-scale mixed farming. To the fisherman also the seaward ends of these valleys have provided shelter from harsh natural, and often harsher human, elements.

The region's lowlands are in general closely confined by abrupt transitions to the upland zones. The major exception is the wide, gentle slope from the eastern fringe of the New Brunswick upland to the Gulf of St Lawrence shoreline, emerging again across Northumberland Strait in the lowland swells of Prince Edward Island. While the confined lowlands, such as the Annapolis Valley of Nova Scotia, the lower Saint John valley in New Brunswick, and the Deer Lake lowland of Newfoundland, share some edaphic and climatic advantages, that of eastern New Brunswick, although the most extensive, suffers from low-nutrient, poorly drained soils.

The sea-floor is an equally varied terrain where topographic variation influences the all-important oceanographic and biotic environments. The floor of the Gulf of St Lawrence is a smooth plain at an average depth of 100–300 feet (30–90 m), divided into a wider western section and less extensive northern and eastern sections by the immense Y-shaped trough of the Laurentian Channel and its tributary Esquiman Channel. This feature extends from near the mouth of the Saguenay River in Quebec for 800 miles (1280 km) to the continental edge south of the Cabot Strait. Steep walls drop from 300 feet along the edge to 1300 feet (480 m) below sea level to a smooth floor ranging from 40–60 miles (64–96 km) in width.

South of Nova Scotia and Newfoundland, the continental shelf varies from 75 miles (120 km) wide off Cape Sable, Nova Scotia, to 225 miles (360 km) wide off Cape Race, Newfoundland. The Scotian shelf is a complex of outer banks at 200–300 feet (60–90 m) nearer the shelf edge, separated from each other in the west by wide basins extending to 500–900 feet (150–270 m) depth, and in the east by a disorganized assemblage of small hills, hollows, ridges, and valleys. The landward shelf area has no extensive banks, being similar in topography to the glaciated bedrock terrain of the neighbouring land. Off southern Newfoundland, Grand Bank is as large as the island itself and is remarkably smooth at about 250 feet (75 m) below sea level. It is separated from smaller banks off the south coast by deep basins and channels that reach 1000 feet (300 m) depth. The shelf off northeastern Newfoundland is restricted to drowned extensions of the larger peninsulas, between which deep channels and basins have been scored in the floors of Notre Dame, Bonavista, and Trinity bays.

Sea floor topography has a dominating influence on the thermal stratification of the sea, and its subsurface movements: factors of great significance for the habitat of marine organisms.

GEOLOGICAL HISTORY

The bedrock, the surficial deposits, and the landforms of the land and sea-floor areas of the Atlantic Provinces are the product of geological events which have occupied approximately the last one billion (10^9) years. Obviously, a detailed account of this history[2] is impossible and inappropriate here. Rather, the purpose of this section is to provide an outline of the major events that were responsible for the character and disposition of earth materials and the landforms developed on them. From the geological map in Figure 1.2, the disposition of rock types can be seen to form the basis of the topographic fragmentation of the region already referred to.

Geologically, the Atlantic Provinces occupy the northeastern extremity of a complex Palaeozoic orogenic belt, the mountainous character of which has been substantially modified by fluvial planation and episodic uplift in the Mesozoic and Cenozoic eras. The continental shelf off southern Nova Scotia and Newfoundland is underlain by a thick wedge of sedimentary rocks, deposited over the southeastern flank of the denuded orogen as the product of that planation. The floor of the Gulf of St Lawrence is a river-worn lowland underlain by the offshore extensions of sedimentary zones of the orogen and, while it lacks a Mesozoic and Tertiary cover, bears a thin veneer of Quaternary deposits of varied origins.

Between late-Precambrian and late-Carboniferous time, approximately 750 to 315 million years ago, the Atlantic region was part of the Appalachian Geosyncline, which occupied a position peripheral to the newly formed Grenville Province around the southern and southeastern edge of the Precambrian Shield. The generally acidic crystalline rocks of the Great Northern Peninsula of Newfoundland have recently been identified as an inlier of that Precambrian orogenic zone, formed some 900 million years ago. The geosynclinal history of the region is, crudely abbreviated, that of the changing importance and shifting position of three depositional-tectonic zones: the relatively calm platform zone near the Shield margin, which grows in areal importance westwards in the St Lawrence Valley and Great Lakes lowlands; the unstable volcanic island arcs and intermittently emergent 'geanticlines' within the geosyncline proper; and the depositional troughs – the shallower miogeosynclines where clastic sediments accumulated closer to emergent lands, and the eugeosynclines with their deeper-water shales and oceanic lavas. These zones, shown in the

2 For information on the geological history of the region the following sources have been extensively consulted: Poole *et al.* 1970; Douglas 1969; Stockwell 1969; Prest *et al.* 1968; Prest 1969, 1970. In this and succeeding sections, major sources consulted are listed in a footnote in order to eliminate repetitive references to them in the text. This convenience is not meant to detract from the fundamental importance of these works to each of the topics discussed.

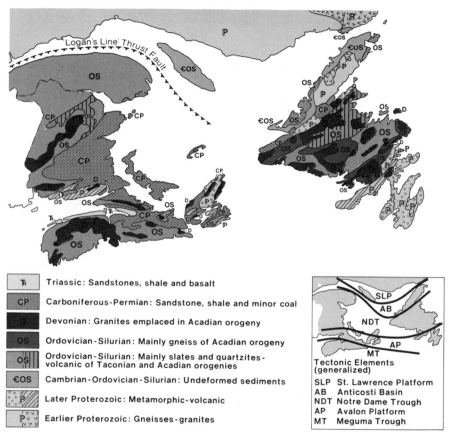

Triassic: Sandstones, shale and basalt

CP   Carboniferous-Permian: Sandstone, shale and minor coal

D   Devonian: Granites emplaced in Acadian orogeny

OS   Ordovician-Silurian: Mainly gneiss of Acadian orogeny

OS   Ordovician-Silurian: Mainly slates and quartzites-volcanic of Taconian and Acadian orogenies

€OS   Cambrian-Ordovician-Silurian: Undeformed sediments

P   Later Proterozoic: Metamorphic-volcanic

P   Earlier Proterozoic: Gneisses-granites

Tectonic Elements
(generalized)

SLP   St. Lawrence Platform
AB   Anticosti Basin
NDT   Notre Dame Trough
AP   Avalon Platform
MT   Meguma Trough

1.2

Bedrock Geology of the Atlantic Provinces

(Source: Generalized after Douglas, 1969 & Poole et al., 1970)

inset to Figure 1.2, are known as the St Lawrence Platform, Notre Dame Trough (with island arcs appearing intermittently), Avalon Platform, and Meguma Trough.

With the opening of the Palaeozoic era, the major tectonic elements were clearly established. In the northwest, clastic and carbonate sediments were laid down on the stable shelf of the St Lawrence Platform through the Cambrian, Ordovician, and Silurian periods, an interval of about 180 million years. These sediments are now exposed undeformed on Anticosti Island and in western Newfoundland, where they were deformed and lightly metamorphosed in places during large-scale gravity-sliding of allochthonous plutonic bodies. To the southeast, in the Notre Dame Trough, great thicknesses of sediments began to accumulate, while over the Avalon Platform shelf deposition of shales and limestones prevailed. In the outermost zone, the Meguma Trough, a thick sequence of sands and muds was deposited by streams and submarine turbidity currents from a landmass to the southeast.

This interval of relative quiescence was closed over much of the geosyncline by the Taconian Orogeny of mid- and late-Ordovician time, about 440 million years ago. The Notre Dame Trough bore the brunt of compressive forces, directed from the southeast, which effected large-scale folding, overthusting, and high-grade metamorphism. Only in the northern part of the Gaspé Peninsula, western and northern Newfoundland, and parts of eastern Newfoundland have Taconian structures not been 'overprinted' by the more widely felt Acadian Orogeny of Devonian times, so that it is difficult to delimit the extent of Taconian disturbance. Nevertheless, it effected enormous horizontal displacement, as along the famous 'Logan's Line' thrust (Figure 1.2). In western Newfoundland, large-scale, northwest-directed gravity sliding of 'klippe' slices emplaced geosynclinal clastic sediments and mafic-ultramafic plutonic bodies upon platform sediments 100 miles or more from their original location, to provide the geological foundations of the spectacular high-relief scenery north and south of Bay of Islands.

The Taconian Orogeny established a more complex pattern of sediment-supplying land areas and depositional troughs, a pattern that controlled the accumulation of thousands of feet of Siluro-Devonian sediments and lavas in the Notre Dame and Meguma troughs. The Acadian Orogeny, occurring through Middle and Late Devonian time, approximately 360 million years ago, deformed these rocks into huge, northeast-trending 'nappes' and effected high-grade regional metamorphism of hitherto unaffected sediments and rocks earlier crumpled by the Taconian Orogeny. In addition, regional plutonism affected the Notre Dame and Meguma troughs so that, with the subsequent unroofing of batholiths, wide areas of central Newfoundland, southwestern Nova Scotia, and north-central New Brunswick are now underlain by Devonian granites (Figure 1.2). The Acadian Orogeny produced structural trends similar to those of the Taconian, but it is difficult to establish whether some are Taconian revivals or Acadian originals. Regardless of their lineage, they give rise to the most striking landforms of the region, such as the fault-line scarps of the Cape Breton highlands and western Newfoundland.

The Acadian Orogeny effectively ended the marine phase of geosynclinal development. In the succeeding Carboniferous and Permian periods the central part of the Appalachian Orogen was occupied by a Basin-and-Range terrain. In the basins a sequence of predominantly continental clastic sediments accumulated in riverine, lacustrine, and swamp environments. Local basin subsidence, upland uplift, and an arid climate were responsible for continual high relief and thick accumulations of reddish strata. During a brief, late Lower Carboniferous marine incursion, chemical precipitates and, locally, coal-forming swamp forests formed, which today provide the basis of coal-mining and salt and gypsum workings in the region.

In upper Carboniferous times, roughly 315 million years ago, this zone of continental deposition became tectonically active in what is termed the 'Maritime Disturbance.' This does not deserve the orogenic status accorded to movements that affected the Appalachian Geosyncline of the eastern seaboard of the United States at this time. In the Atlantic region, weaknesses in the Acadian basement were re-activated and were reflected in Carboniferous strata as folds, commonly with near-vertical limbs, and normal faults.

The Mesozoic era opened on a 'continental' scene in the Atlantic Provinces, contrasting with the predominantly 'maritime' aspect which had prevailed for most of the Palaeozoic. Triassic red-beds around the Bay of Fundy represent continued terrestrial sedimentation in a dry climate. But it is the spectacular Triassic lavas of the Fundy area which speak of major changes in the regional geological sequence. These flood basalts were associated with the early phases in the most recent opening of the Atlantic Ocean.

Several recent studies of Mesozoic and Cenozoic rocks on the continental shelf off Nova Scotia and Newfoundland have contributed to an emerging picture of post-Triassic events in the Atlantic region. Conclusions drawn from analyses of samples dredged from the sea floor (King and MacLean 1970), from seismic reflections profiles (King et al. 1970), and from drill cores (Bartlett and Smith 1971), converge to give a picture of the shelf as an area across which the sea repeatedly transgressed during periods of stability, and from which it often withdrew in response to local differential uplift. Subaerial and near-shore depositional environments prevailed throughout. Present land areas appear not to have been overlapped by the sea in Mesozoic and Tertiary times.

These conclusions bear upon the manner in which the preglacial landforms of the region have evolved. Traditionally, hypotheses of landform evolution have been based upon the Davisian notions of peneplanation and progressive adjustment of drainage patterns to geological structure, interrupted by relatively short periods of rejuvenating uplift. Workers such as Goldthwait (1924) in Nova Scotia, Alcock (1928, 1935) in Gaspé and northern New Brunswick, and Twenhofel (1912; with Mac-Clintock 1940) in Newfoundland saw the strikingly level upland skylines and gently sloping upland valley floors as remnants of peneplains which had developed through Mesozoic and Tertiary time. The southeastward and eastward slope of an envelope surface over the uplands, and the general southerly alignment of the few river segments which transect geological structures, were interpreted as vestiges of an original surface upon which 'consequent' drainage lines were initiated.

Closer examination of parts of the Atlantic uplands (Bird 1964; Brookes 1964) has revealed that, in detail, their surfaces are made up of many small planation surface remnants bevelling geological structures. These bevels commonly slope through a height range of 100–200 feet (30–60 m) and are separated by vertical intervals of the same order. Their form and pattern speak of a fluvial origin.

This more detailed interpretation of upland surface form accords more satisfactorily with the evidence of crustal instability from the continental shelf. It is tentatively put forward that uplands have undergone continuous subaerial denudation, accompanied by intermittent small uplifts, and that there is no evidence in the form of the uplands to indicate that baselevelling was ever accomplished.

Such a scheme can hardly accommodate an original surface drained by southeastward and eastward-flowing 'consequent' rivers. Rather, it leads to an hypothesis, corollary to that offered above, that major drainage lines have persistently followed lithological and structural weaknesses towards a coastline on the continental shelf and that uplands have just as persistently reflected stronger rock zones. With intermittent uplift, reflected in upland surface form and continental shelf rock sequences, transection of geological structures would likely arise from the greater vigour of upland stream segments, which were able to maintain such courses against the tendency to follow structural weaknesses.

The record of the Quaternary period in the Atlantic Provinces is limited to the Wisconsinan stage and predominantly the final, or 'Classical Wisconsin,' substages, occupying the last 18,000–20,000 years. In that interval glacial erosion and deposition have merely put 'finishing touches' to the major landforms of the region. They have had far more important effects upon soil parent materials. Further, the encroachment of the sea upon the land that accompanied the melting of the ice sheets has radically affected the pattern of land and sea distribution.

Investigations of glacial phenomena in the region have been dominated by a debate over the relative extent of an ice sheet from the Laurentide centre of glacial outflow and ice from local outflow centres in Appalachian Canada. Opinion has swung from assigning important roles to local ice outflow centres in highlands such as those of Newfoundland, the Gaspé Peninsula, northern New Brunswick, and Cape Breton Island – an opinion that dominated the late-nineteenth and first decades of the twentieth century – to the notion of overwhelming Laurentide ice at the glacial maximum, with only late-glacial independence for local ice caps in highlands, an hypothesis favoured since the 1940s and 1950s. Currently, some evidence is promoting a return to the earlier notion (Prest and Grant 1969; Brookes 1970).

The island of Newfoundland offers the clearest evidence for an independent ice-cap at the last glacial maximum, in the orientation of striae and flow-moulded bedrock and drift materials (Figure 1.3). In the mainland provinces and Prince Edward Island such features do not allow firm conclusions to be drawn, since their orientation alone can often be attributed to either Laurentide or local ice. Features diverging from the common southeast trend are most often the result of late-glacial re-orientations of ice margins: they say little of full-glacial flow patterns. Erratics have, however, proved conclusive in relating the maximum glaciation of western Newfoundland to an island ice cap, and in Prince Edward Island foreign boulders along the north shore are assigned a northwesterly or westerly provenance. Erratics of northern provenance over the Notre Dame Mountains of Gaspé must have been emplaced by Laurentide ice, but they offer no evidence for the overriding of the entire Maritime Provinces (Prest 1970).

Glacial erosion achieved its most spectacular effects in coursing down valleys incised into the margin of the Newfoundland plateau block. Along the coasts overdeepened valleys, drowned since deglaciation, attain depths of 1000 feet (300 m) below sea level, with steep walls rising 1000–2000 feet (300–600 m) above the sea on each side. On the shelf off the eastern half of the island, glacial erosion has formed basins that attain 1000–2000 feet depth, as in Fortune, Conception, and Trinity bays and the arms of Notre Dame Bay. The Laurentian Channel has also been identified as a glacially modified river valley (King 1970). Prest and Grant (1969) have drawn attention to the possibility that the Channel diverted ice flowing from the Laurentide centre of dispersal, thereby allowing local highland-centred ice caps to expand over the Maritime Provinces.

The highlands of New Brunswick and Nova Scotia, on the other hand,

→   Striation, ice flow known ice flow direction known
////   Features molded by ice flow
      Ribbed moraine
●40   Marine limit elevation (metres)
13.5   Approximate age of marine limit x 10^3 years B.P.
⊥⊥⊥⊥⊥⊥   End moraines (hatching on proximal side)
———   Line of zero postglacial emergence
      Maximum marine overlap

1.3

Glacial Map of the Atlantic Provinces
(Source: Generalized after Prest et al., 1968 with additions)

bear surprisingly little evidence of glacial modification. Plateau surfaces
in their smooth appearance bear evidence of the fluvial planation discussed
earlier. Valleys dissecting their margins show only occasional signs of the
overdeeping and oversteepening so common in Newfoundland.

Unspectacular to the eye, but far greater in volume, has been the glacial
erosion achieved in the production of the almost continuous and, in the
lowlands, generally thick, till mantle of the Maritime Provinces. More
will be said of till lithology in connection with the soils of the region, but
it may be noted here that, despite the common presence of far-travelled

erratics, the bulk of the till is of local derivation and reflects local bedrock lithology closely. Thus, it is possible to distinguish between the bouldery, sandy loam tills of the crystalline uplands, and the generally boulder-free sandy loams of the extensive sedimentary terrane of eastern New Brunswick, Prince Edward Island, and northern Nova Scotia. Over the upland of peninsular Nova Scotia, the extensive granite and quartzite terrane yielded a thin, bouldery, sandy till, while deeper and more loamy tills are locally prevalent over the associated slate belts. Notable drumlin fields occur west and northwest of Halifax, and at the western extremity of the peninsula.

Of the 70 per cent of Newfoundland not classified as bedrock outcrop, thin till, or organic terrain, most is covered by thicker, bouldery and sandy till and coarse, ice-contact stratified drift. Only in isolated lowlands on the west coast of the island does the drift acquire a more loamy texture, but boulders are still common.

From terminal positions on the continental shelf, attained approximately 18,000 years ago, the ice sheets were calved back rapidly by the eustatically rising waters of the Atlantic Ocean, especially along the axes of offshore troughs such as the Laurentian and Esquiman channels, the Bay of Fundy, and those off eastern and southern Newfoundland. The radiocarbon dates shown in Figure 1.4 give a crude idea of the progress of marine overlap.

The encroachment of the sea in calving bays isolated local ice caps in highland areas such as the upland of peninsular Nova Scotia, Cape Breton Island, the Gaspé Peninsula, northern New Brunswick, and the central plateau of Newfoundland, about 11,000 to 12,000 years ago. Ice flow directions then became more variable because of the more diverse orientations of glacier margins. It is largely these final flow patterns that are recorded in present-day features such as striae, flow-moulded surficial deposits, and the patchy recessional moraines. Only in western Nova Scotia has a reactivation of a local ice cap been documented. There, the presence of granitic erratics of southern provenance atop the basaltic North Mountain demands northward flow from South Mountain, probably caused by a steep glacier gradient into the Bay of Fundy while marine inundation was in progress.

By 10,000 years ago, the Atlantic Provinces were almost free of glacier ice. Prest (1969) shows a small remnant of the Newfoundland glacier on the eastern slope of the Great Northern Peninsula of Newfoundland, and it is likely that cirque glaciers also remained in these highlands and those of the Gaspé Peninsula.

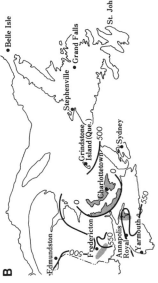

B

Belle Isle

Grand Falls
Stephenville
St. John's
Grindstone Island (Que.)
Charlottetown 500
Sydney
Edmundston 500
Fredericton 550
Annapolis Royal
Yarmouth 550

Areas with annual water deficit (100 mm field capacity)

Mean Annual Potential Evapotranspiration (millimetres)
(Source: Atlas of Canada, 1969)

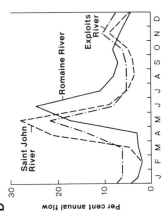

D

Per cent annual flow

30

20

10

0

J F M A M J J A S O N D

Romaine River
Exploits River
Saint John River

Mean Annual Streamflow Distribution
(Source: Hydrological Atlas of Canada, 1969)

Romaine River, Quebec;
15 kms from mouth;
mean annual flow 297 c.m.s.

Exploits River, Nfld;
at Grand Falls;
mean annual flow 219 c.m.s.

Saint John River, N.B.;
at Pokiok;
mean annual flow 728 c.m.s.

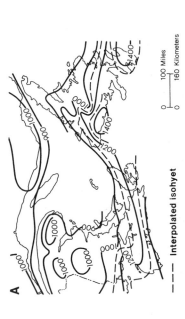

A

1000
1400
1000
1200
1000
1000
1000
1000

— — — Interpolated isohyet

0   100 Miles
0   160 Kilometers

Mean Annual Amount of Precipitation (millimetres)
(Source: Canada Department of Transport)

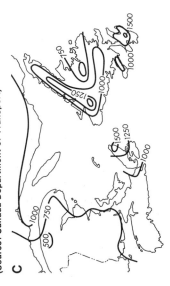

C

750
1250
1000
1500
1000
500
1250
1000
1000
750
500

Mean Annual Runoff (millimetres)
(Source: Hydrological Atlas of Canada, 1969)

1.4

Hydrologic Elements of the Atlantic Provinces

CLIMATE AND HYDROLOGY

The climate of the Atlantic Provinces reflects the influence of many factors operating at several scales. A position between 43° and 51° North latitude gives rise to a marked seasonality of insolation and temperatures. This latitudinal spread is sufficient to place Yarmouth, N.S., within the Köppen 'C' climates, and Belle Isle, Newfoundland, almost within the 'E' type (see Table 1.1). The same factor sets the region astride the most travelled paths of east-moving, mid-latitude, low-pressure cells. A situation on the eastern side of the continent further induces strong seasonal contrasts in the character and movements of air masses. The cold Labrador ocean current washing the eastern fringe and the warm Gulf Stream along the southern extremity play significant roles in air mass modification, and a similar significance attaches to ice-covered and ice-free sea surfaces in winter. On land areas, winter snow and ice covers contribute much to negative heat budgets. Air mass flows are generally unhindered by topographic barriers and corridors, although important local influences prevail.

In Table 1.1 monthly temperature and precipitation data are given for eleven stations, selected to permit several situations to be compared: northern and southern, eastern and western, marine, coastal, and interior, winter ice-bound coastal and ice-free coastal, cold and warmer offshore waters in summer.[3] For the stations listed, a mean January temperature of 20°F (−6.6°C) and a mean July temperature of 62°F (16.7°C) reflect the seasonal contrast in air mass flows. In winter, air mass dominance is shared between continental arctic air from the northwest and maritime polar air from the northeast and north. Summer air masses are of maritime tropical character, flowing into the region from the western Atlantic Ocean and the Gulf of Mexico in a general northeasterly direction. The annual regime of air mass flows may be understood by reference to prevailing synoptic situations. The mean position, extent, and trajectories of pressure centres in mid-latitude North America produce a reasonably reliable contrast between continental interior and eastern maritime synoptic situations.

With a well-developed, slow-moving, high-pressure cell over southern Manitoba and the northern prairie States in winter, producing sub-zero temperatures, clear, stable air, and light winds, the Atlantic Provinces will normally be under the influence of a northeast-moving low pressure cell. In advance of this, cold, southeast winds bring snow and temperatures in the mid-20s to low 30s (F°). As the warm sector of the system moves

---

3   Stations are located in Figure 1.4(b).

through the region, a 12–24-hour interval of above-freezing temperatures and rain or wet snow will be followed by strong northerlies in the rear of the cold front. Over Newfoundland these winds will be more from the northeast and they bring the worst weather conditions the island experiences. Temperatures in the twenties, strong winds, and very heavy snowfall commonly cause major disruptions of internal and external communications. Over the Maritime Provinces, airflow behind the cold front will be more northwesterly and, although some snowfall can usually be expected, this air is more the herald of stable, clear, cold weather. These more settled conditions are never as long-lasting as they are in the continental interior: air mass trajectories are too variable to allow a high-pressure cell to stay for longer than one or two days. Maritime polar air off the northwest Atlantic, continentally modified North Pacific air from the interior, and maritime tropical air from the southwest – all consort to hasten the exodus of inherently sluggish high-pressure cells.

Winter temperatures reflect the differences between marine, coastal, and interior locations. For example, in Table 1.1, compare the January mean at Edmundston, N.B., with that of Belle Isle, Nfld. – the latter is 5° farther north, yet its annual minimum temperature is 4°F warmer. Edmundston and St John's, Nfld., are at a comparable latitude, yet the St John's minimum in February is 14°F warmer. In Nova Scotia, the 5°F difference in February between Kentville inland and Yarmouth on the coast is another example of maritime influence. Winter minima at coastal localities also differ according to the presence or absence of sea-ice. For example, Charlottetown, P.E.I., experiences sea-ice for one to three winter months and its mid-winter temperatures are 2–3°F cooler than those at ice-free Sydney, N.S.

With rising net radiation values in spring, incursions of maritime tropical air into the region become increasingly frequent until, in late summer, the southern border of arctic and polar air masses has migrated northwards to lie diagonally across the centre of Quebec–Labrador. The pressure gradient between the Bermudian high and Icelandic low is halved, reducing the frequency of northeasterlies. Low-pressure cells from the Canadian northwest pass to the north of the area in summer and are replaced by those born in the central United States of the conflict between tropical Gulf and modified Pacific air. While these produce rain in their passage over the Atlantic Provinces, they are less intense than winter storms.

Proximity to the sea is reflected in lower summer temperatures and a delay of the warmest month to August. In Table 1.1, for example, compare Yarmouth, N.S., in August (62.4°F) with Kentville in July (66.6°F);

Table 1.1 Mean monthly temperature (°F) and precipitation (inches) for selected stations in the Atlantic Provinces*

	Jan.	Feb.	Mar.	April	May	June	July	Aug.	Sept.	Oct.	Nov.	Dec.	Year	
Grindstone Island, Magdalen Is., Que.: Central Marine														
	20.7	18.7	24.2	32.8	42.4	52.0	61.8	62.5	55.8	46.0	36.6	26.6	40.1	43.8[a] temperature
	3.5	2.72	2.37	2.29	2.77	2.49	2.37	3.07	3.35	3.40	3.87	3.66	35.88	1.58[b] precipitation
	24.2	19.5	14.5	7.2	0.5	T	0	0	T	0.4	5.3	18.8	90.4	37%[c] snowfall
Edmundston, New Brunswick: Northwest Interior														
	9.5	11.6	22.5	36.7	50.2	59.8	65.1	62.7	54.2	43.5	30.5	15.3	38.5	54.6
	2.99	2.83	2.41	2.45	3.12	4.24	4.01	3.66	3.44	3.35	3.30	2.85	38.65	1.83
	27.4	27.0	20.1	9.0	0.7	0	0	0.1	0.1	2.6	13.9	21.9	122.7	52.5%
Fredericton CDA, New Brunswick: West Interior														
	15.5	16.6	26.9	39.8	51.5	60.4	66.5	65.0	56.8	46.0	34.6	19.5	41.6	51.0
	3.91	3.25	3.20	3.47	3.38	3.78	3.49	3.19	3.66	3.88	4.34	3.65	43.15	1.14
	22.6	21.2	15.9	1.0	0.3	0	0	0	1.1	1.1	7.6	16.5	92.2	31.6%
Yarmouth A, Nova Scotia: South Coastal														
	27.7	27.3	32.6	40.8	49.1	56.4	61.9	62.4	57.5	49.8	41.8	31.8	44.9	35.1
	5.86	4.58	4.35	3.94	3.82	3.31	2.89	3.29	3.80	4.19	4.95	5.02	50.0	2.97
	23.2	20.3	14.1	2.6	T	0	0	0	T	0.1	3.1	18.3	81.7	25%
Kentville CDA, Nova Scotia: South Interior														
	22.5	22.2	29.6	40.4	50.9	59.7	66.6	65.3	58.3	48.1	39.0	27.0	44.1	44.4
	4.69	3.66	3.62	2.78	2.88	2.87	2.54	3.52	3.78	3.62	4.41	4.14	42.51	2.15
	23.0	20.2	17.3	4.2	0.1	0	0	0	0	0.3	4.5	17.7	87.3	29.6%
Sydney A, Nova Scotia: Central Coastal														
	24.3	22.3	27.9	37.0	46.7	55.9	64.9	64.9	58.0	48.3	39.5	29.1	43.2	42.6
	5.13	4.67	4.42	3.87	3.68	3.50	2.78	4.04	4.08	4.54	5.62	5.04	51.37	2.84
	21.8	23.6	19.5	8.3	0.7	0	0	0	T	0.3	3.5	17.6	95.5	26.6%

Table 1.1 (Concluded)

	Jan.	Feb.	Mar.	April	May	June	July	Aug.	Sept.	Oct.	Nov.	Dec.	Year			
Charlottetown CDA, Prince Edward Island: Central Coastal														46.9	2.52	34.8%
	20.5	19.9	27.3	37.9	49.0	58.3	66.8	66.1	58.6	48.3	38.2	25.8	43.1			
	4.15	3.39	3.22	2.95	3.13	2.91	3.07	3.41	3.94	3.92	4.43	4.22	42.74			
	26.0	24.4	17.2	7.0	0.2	0	0	0	0	0.6	5.1	21.9	102.4			
Belle Isle, Newfoundland: North Marine														38.1	1.84	30.0%
	13.5	13.9	19.8	27.6	33.9	41.3	49.1	51.6	45.7	37.3	28.5	19.5	31.8			
	1.75	2.12	2.18	2.23	2.54	3.55	2.84	3.44	3.59	3.37	3.32	2.63	33.56			
	12.9	15.0	15.6	11.7	2.9	0.7	0	0	0.6	3.3	10.3	19.0	92.0			
Stephenville A, Newfoundland: Central Coastal														40.1	2.35	35.0%
	23.0	21.7	27.0	35.2	45.0	54.0	61.5	61.8	54.9	45.0	36.5	28.1	41.2			
	4.01	3.70	2.12	2.02	2.81	3.03	3.39	3.32	4.18	3.88	4.37	3.79	40.62			
	25.8	23.2	12.9	7.5	0.7	0.1	0	0	T	0.9	5.1	24.2	100.4			
Buchans, Newfoundland: East Interior														44.3	2.01	40.0%
	16.9	15.7	21.0	31.5	42.1	52.6	60.0	59.0	51.8	41.5	32.5	22.2	37.3			
	3.31	2.92	2.35	2.10	2.42	2.60	2.94	3.68	3.57	3.33	4.11	3.40	36.73			
	23.2	21.0	18.1	9.6	1.3	0.9	0	0	T	1.7	9.9	20.6	106.3			
St. John's West CDA, Newfoundland: East Coastal														36.3	4.43	29.6%
	25.0	23.6	26.9	34.5	42.3	50.4	59.5	59.9	53.0	44.0	37.6	28.8	40.5			
	6.85	7.07	6.32	5.18	4.39	3.20	2.89	4.02	4.07	5.40	7.32	7.08	63.79			
	25.8	23.2	12.9	7.5	0.7	0.1	0	0	T	0.9	5.1	24.2	100.4			

*Source: Canada, Department of Transport 1968a, 1968b.
a Range of mean monthly temperatures.
b Range of mean monthly precipitation.
c Percentage of snow-season precipitation received as snow (reduced to water equivalent of 0.1).
A Airport recording station.
CDA Canada, Department of Agriculture recording station.

and Grindstone Island in August (62.5°F) with Edmundston in July (65.1°F); and Stephenville, Nfld., in August (61.8°F) with Buchans in July (60.0°F).

Summer air temperatures at coastal stations are further affected by off-shore water temperatures. Compare Stephenville and St John's, Nfld., in August. Even the 1–2°F differences in July between Charlottetown and Sydney might be partly accounted for by the warmth of Northumberland Strait waters compared with those of the open Atlantic off Cape Breton Island.

Mean annual precipitation at the stations selected for Table 1.1 varies from about 33 inches to about 64 inches (84–160 cm). At most stations this is quite evenly spread over the year. Fredericton, N.B., shows the least annual variation (1.14 inches), and St John's, Nfld., the greatest (4.43 inches).

The timing of precipitation maxima and minima is quite variable at the stations listed in Table 1.1. Table 1.2 shows the number of stations, of the eleven shown, at which each month is one of the four driest and one of the four wettest months of the year. November, December, and January most frequently appear among the four wettest months. This reflects the heavy precipitation associated with winter storms. It is worth noting that, at stations where these months are the wettest, the water equivalent of the snow recorded is only 30–40 per cent of the precipitation recorded in snowy months (applying a factor of 0.1 to measured snow amounts to arrive at water equivalent). This might well be too low for this region, however. Edmundston, N.B., and Belle Isle, Nfld., are the only two stations where none of these months is one of the four wettest. A 'continental' pre-cipitation regime explains Edmundston's case, whereas Belle Isle, a north-ern location, is fully under the influence of dry continental arctic air in mid-winter and is also icebound.

Late-summer and autumn low-pressure cells, generated as tropical storms and hurricanes in the tropical western Atlantic, bring heavy rains to parts of the region to explain the fact that August, September, and Oc-tober rate second among the four wettest months of the year (Table 1.2). June is one of the four wettest months only at Edmundston, N.B., with its continental regime, and at Belle Isle, Nfld., within the polar frontal zone at that time.

Spring and early summer are notably dry over the entire region. This is fortunate, for, with soils at or close to field capacity throughout the winter, rivers would rise catastrophically if the snow-melt runoff they received were to be augmented by heavy thaw-season rains.

Spring and summer are the foggiest seasons. Summer is the season of

Table 1.2 Driest and wettest months at eleven stations in Table 1.1[1]

	Jan.	Feb.	Mar.	April	May	June	July	Aug.	Sept.	Oct.	Nov.	Dec.
One of 4 wettest Months (maximum 11)	8	4	0	0	0	3	1	3	4	5	9	7
One of 4 driest Months (maximum 11)	1	3	6	8	6	8	6	3	2	0	0	1

1 Figures show the number of stations, of the 11 listed in Table 1.1, at which the month in question is one of the 4 wettest, and one of the 4 driest, of the year.

greatest contrast between sea-surface and overlying air temperatures. An average of more than one in four fog days occurs in March, April, and May along the Fundy littoral, Cape Breton Island, Newfoundland east of a line from Placentia Bay to Notre Dame Bay, and in the central Gulf region as far north as southern Anticosti Island. In June, July, and August this frequency expands to cover all of Nova Scotia, southern New Brunswick, the entire Gulf, and coastal Newfoundland.

The incidence of coastal fog is strongly related to antecedent weather conditions. For example, a warm, onshore, southwest wind on a west-facing coast will drive warm water inshore, so that the saturation deficit of air in contact with it is small. An offshore wind will then cause this warm water to be replaced by upwelling cold water over a wide littoral zone. If, then, the southwest wind returns, it will pass over cold inshore water and the high saturation deficit will give rise to dense, thick fog which will persist until the wind veers.

The influence of atmospheric elements upon the heat and water budgets of different areas of the Atlantic Provinces has not received the detailed attention afforded to it in areas of higher agricultural capability. While crude estimates of soil moisture deficit, based upon a four-inch (10-cm) field capacity assumption, show that only small coastal lowland areas experience a deficit in the average year (Figure 1.4(b)), values of field capacity closer to one inch (2.5 cm) commonly prevail in sandy and gravelly soils. Soil droughts can thus affect sensitive tree- and ground-fruits and vegetable crops in the thermally most suitable areas of the region.

Industrial, power, and urban development require that the area's water resources be known and predictable so that planning can take account of source areas, water quality, and storage variability. However, knowledge is inadequate at present concerning the magnitude of water budget elements – precipitation, runoff, evapotranspiration, and storage. The maps shown in Figure 1.4 are meant to illustrate the discordance between various sources of data. Note, for instance, that, while precipitation over northern Cape Breton Island is 40 inches (100 cm) and potential evapotranspiration is 20 inches (50 cm), the runoff is shown to be 60 inches (150 cm). It is beginning to be realized that in snow climates there is an underestimation of precipitation because of erroneous assumptions about patterns of snow and ice accumulation and density. Other errors and unknowns include the unmeasured contribution to soil moisture from condensation in coastal and highland areas; false assumptions about winter evapotranspiration of evergreen conifers; unmeasured winter evaporation of water from snow and ice; and overestimates of evapotranspiration where non-vascular plants cover wide areas, as in Newfoundland.

Streamflow is the easiest hydrologic element to measure, and it is used to check the accuracy of estimates of precipitation/evapotranspiration. The annual distribution of streamflow (Figure 1.4(d)) shows a May peak of between 20 and 30 per cent of annual flow which is clearly the result of snow and ice melting. A rapid decline to an August minimum of 5 per cent reflects rising evapotranspiration, especially in southern and inland areas, whereas in coastal areas a slight summer minimum of precipitation also contributes. A secondary runoff peak in November, amounting to 10 per cent of annual flow, results from late-autumn rains and lowered evapotranspiration.

Several water-supply problems are of immediate concern in the Atlantic Provinces. The dispersed rural population has traditionally relied upon shallow wells for domestic water supplies. With increasing urbanization and nucleation around new industrial and power developments, reservoirs will supply more and more domestic water. Hydrologic and ecologic planning and sound design will be required to ensure the reliability of a good quality supply, with the subsidiary benefits of recreational use and nature conservation.

THE MARINE ENVIRONMENT[4]

Meteorologic, oceanographic, and bathymetric elements interact to impart characteristics to sea water which vary over space and time to produce environmental patterns and rhythms. These, in turn, influence marine biota. This section outlines the salient features of the marine environment and, as an example, illustrates the influence these have on an organism that has always been important to the people of the region – the Atlantic cod.

The more important oceanographic elements are 'autochthonous' rather than transported from another area, comprising less extensive, less persistent, and shallower drifts of water. In some areas they are related to the intrusive Labrador Current; in others, they are local and directed more by the configuration of coastlines and bottom topography than by prevailing winds. Thermal stratification is influenced by local factors as well as by higher-order controls (Figure 1.6). Three oceanographic regions may be recognized: Gulf of St Lawrence water, shelf water off the south

4   Sources consulted: Black 1959–63; Forward 1954; Hachey 1961; Leim and Scott 1966; Templeman 1966.

```
──────▶   Persistent
------▶   Variable
├──────┤  Line of section (see 1.6)
```

1.5

Surface Circulation of Atlantic Region Offshore Waters in Spring and Summer

coasts of Nova Scotia and Newfoundland, and shelf water off eastern and northern Newfoundland.

The Gulf of St Lawrence, as an almost landlocked sea, experiences greater seasonal changes in current vectors and thermal stratification than open shelf seas. Surface circulation within it is counterclockwise (Figure 1.5). Deeper movements are confined to the Laurentian Channel, in depths from 300 to 1300 feet (*ca.* 100 to 400 m) – into the Gulf in summer and out in winter, through the Cabot Strait. The thermal structure in winter comprises two layers. With declining net radiation after July, surface water cools and, when it reaches the temperature of maximum density, 39.2°F (4°C), it sinks. By December, the water over shallow shelf areas is entirely below this temperature. When the surface water reaches 29°F

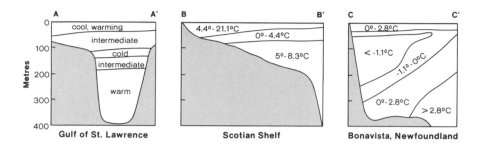

1.6

Thermal Structure of Atlantic Offshore Waters in Summer
(Source: Hachey, 1961 & Templeman, 1966)

(−1.7°C) it begins to freeze. In the Laurentian and Esquiman channels, this cold-water layer lies over warmer, isothermal water slowly moving out through the Cabot Strait (Figure 1.6). Sea ice conditions will be discussed later.

In summer, surface waters warm to about 68°F (16°C) in the southwestern Gulf and to about 55°F (13°C) in the northeast and northwest. This warming layer gradually thickens until, in late summer, it has trapped the remnants of the winter cold water between it and the shallow sea floor or warmer water in the Laurentian Channel (Figure 1.6). Surface salinities are reduced in summer by land runoff, and in shallower inshore areas warm, brackish water may extend to the bottom. Summer and autumn storms play an important role in determining the depth of inshore mixing between warm surface water and the colder bottom layer.

Shelf water off southern Nova Scotia and southern Newfoundland is highly stratified. In summer, an upper layer at 40–80°F (4.5–26°C), of low to moderate salinity, attains a maximum thickness of 240 feet (73 m), so that it is bottom water in inshore areas (Figure 1.6). An intermediate layer at 32–39°F (0–4°C) and higher salinity, formed during the winter, lies between 120 and 500 feet (36–150 m) and is, therefore, characteristic of much bottom water on the outer shelf in summer. It is derived by horizontal transport from the northeast and is, thus, related to the Labrador Current. In the isolated basins between each bank, at 300–600 feet (90–180 m), bottom water in summer is warmer, with a temperature of 41–46°F (5–8°C), and is probably derived from the continental slope.

The northeast and east Newfoundland shelves are under the influence of the Labrador Current throughout the year. Inshore waters are colder in winter and less saline than the axial part of the current. In summer, inshore warming reverses the temperature gradient, while runoff strengthens that of salinity. Summer surface temperatures range from 52–56°F (11–13°C) inshore, to 43–45°F (6–7°C) offshore (Figure 1.6), but the storminess of this area causes rapid changes in the extent of warmer inshore water and the depths at which cooler water may be found – changes of vital significance to the fishery.

The relations between environment, biota, and man can be illustrated with reference to the Atlantic cod (*Gadus morhua*) and its human predators. The adult cod is a gregarious, cold-water groundfish with a varied diet. Its habitat is restricted to a relatively narrow temperature range and is limited further by high sunlight penetration and low food supply, limits that are in turn dependent upon a complex of elements.

Cod spawning grounds and seasonal migrations are shown in Figure 1.7. Spawning occurs in the spring months in most areas, in bottom waters near 41°F (5°C). These waters are at depths of 600 feet (180 m) off northeastern Newfoundland where 'banks' are small and where too-cold Labrador Current water is at the surface. Over the Grand Banks there is a simliar stratification in spring, when bottom water is in the appropriate temperature range for spawning cod. Throughout the summer, surface water warms into a range suitable for cod following food supplies inshore. Near the shelf edge, warm continental slope water derived from the Gulf Stream makes local incursions over the sea bottom, so that cod are restricted to a higher layer.

On the Scotian shelf, cod spawn in April in water which extends to the bottom at 100–250 feet (30–80 m), beneath westerly-drifting cooler water derived from the Labrador Current. This will itself warm through the summer months to temperatures above the range of cod. In the Gulf of St Lawrence, the cod range is similarly restricted to the thin layer of cold water sandwiched between the summer-warming layer and the sea floor or warmer 'Channel' water. In spring, surface warming has not progressed so deep that cod cannot spawn over shelf areas in the southern and western Gulf.

The five million or so eggs laid by each female cod rise to warmer surface waters and hatch into pelagic larvae about 4 mm long. For two weeks these larvae feed on plankton and grow to 5 mm, after which they sink to enter the 'fry' stage in waters that are cooler. After several months they have grown to about 4 cm and are 10–20 cm long after one year. Then, as adults, they participate in the migration that is repeated each of the remaining 15 to 20 years of their life.

Climatic and oceanographic elements, as well as cod movements, restrict the local fishery to the summer months: 75 per cent of landings are made in June, July, and August. The success of the inshore (i.e. small-

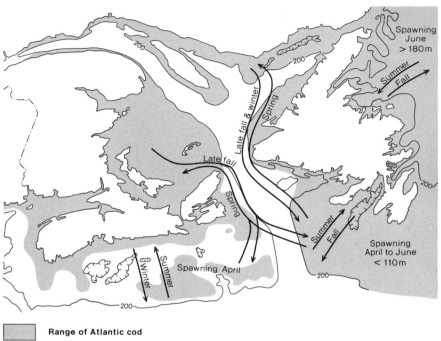

Range of Atlantic cod

Fall ➤ Migration direction and season

1.7
Range and Migrations of Atlantic Cod in the Atlantic Ocean

boat) fishery is influenced by many physical factors, some of which can be considered. As the surface layer warms in the early summer, cod rise off the bottom and begin to feed on the smelt-like capelin which they follow on its spawning trip to the shallows. If the weather is calm, this warm layer will remain relatively thin and cod will be available only over a narrow inshore zone. With continuing calm, the surface water, with its addition of fresh runoff, will warm and mix with difficulty with underlying colder water. Cod will then be available over a widening inshore zone to handline fishing and shallow traps through July and early August.

If onshore winds create a heavy sea in the early summer, mixing of the warm surface and deeper, colder waters creates a water layer suitable for cod through 120–180 feet (36–54 m) – two or three times its calm-weather thickness. Under these conditions a wider bottom area becomes available for cod browsing and the schools do not have to follow the capelin inshore to the same extent. They then become available to deeper-water traps, line trawls, long lines, and gill nets.

If offshore winds prevail in early summer, the warming surface water is driven out to sea and is replaced by upwelling, cold, bottom water that is unfavourable to cod. The fish may then seek out the deepest warm water (continental slope water along the southern shelves) or move offshore with the wind-driven warmer surface water. This latter condition favours the single fisherman 'jigging' from a small boat (often only an 18-foot dory with an outboard motor), since he can make a day's trip some five miles offshore and back over a smooth sea and land many cod from relatively shallow depths.

The influence of water transparency can work to affect cod distribution in an opposite sense to temperature. The mature cod is intolerant of high sunlight penetration. Under calm weather, clear-sky conditions, when cod would normally move inshore within a thin surface layer of habitable water, light penetration will be high in the shallows before planktonic blooms have clouded the waters. The cod will then be forced to seek alternate habitats – a difficult task if deeper waters are too cold.

This brief discussion has oversimplified many matters and omitted more but has, I hope, served to illustrate the importance and complexity of environmental influences on a marine food resource.

The growth of sea-ice in the region is one element of the marine environment that restricts the cod fishery but, on the other hand, it has served as the dominant influence on the location and success of another fishery that has been of traditional importance to the region – the seal fishery.

Sea water begins to freeze at 29°F (-1.7°C) and thickens at the base as heat energy in the water is given up to the air. Ice will form first in inshore areas, where water is less saline and the surface is less disturbed by waves. Once formed, it is driven by wind at a rate estimated at 4 per cent of the wind velocity from any direction, so that, theoretically, ice-drift can be estimated as a resultant of wind vectors. Strong winds of short duration are more effective than persistent breezes in moving sea-ice since they exert greater stresses on the ice surface, and can thus move ice against currents. Ice distribution, therefore, may be highly variable in time, especially in an enclosed body such as the Gulf of St Lawrence.

The Gulf receives ice from the St Lawrence River, where it forms early, and from the Arctic via the Labrador Current and the Strait of Belle Isle, in addition to local ice (Figure 1.8). St Lawrence River ice is driven into the open Gulf by the Gaspé Current, assisted by strong northwest winds which follow the passage of cold fronts. Northern ice rarely reaches mid-Gulf from the Strait of Belle Isle area against the dominant northwesterly

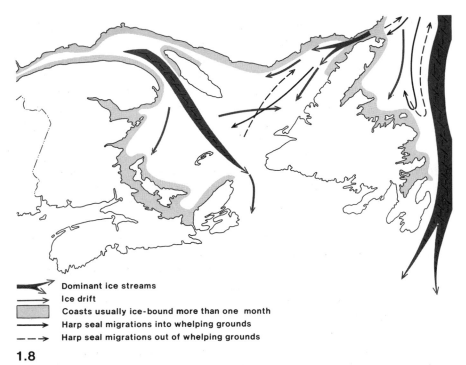

Dominant ice streams
Ice drift
Coasts usually ice-bound more than one month
Harp seal migrations into whelping grounds
Harp seal migrations out of whelping grounds

1.8

Sea-Ice Movements & Harp Seal Migrations in the Atlantic Region

winds, but is restricted to the Quebec–Labrador and northwest Newfoundland shores.[5] It is difficult to speak of average ice conditions in the Gulf of St Lawrence because of the variation of climatic conditions within and between winters. However, 'heavy' and 'light' conditions can be identified and related to specific influences. In a 'heavy' ice winter, the lower St Lawrence freezes over and contributes floes to the Gulf by mid-December, and the north and west shores of the Gulf will be frozen two weeks later. Winds from the west, northwest, and north will drive this ice across the Gulf and southeastwards to jam the Cabot Strait by late January. A persistent ice-free current flowing into the Gulf around Cape Ray, Newfoundland, keeps that shore open in most years, but when strong southwest

5   However, it is interesting to note that the subject of this ice enters into the folk-culture of Cape Breton Island, as in this extract from a modern Gaelic poem by N.K. MacLeod: 'If the Straits of Belle Isle, nine miles across, were closed with an enormous pile of stones, there would be warmth in March everywhere south of it; the ice of Baffin's Bay would go past without harm.' (Trans. C.W. Dunn, 1953)

winds drive ice against it, the Cabot Strait then becomes so blocked that mariners refer to the condition as 'The Bridge.'

In 'light' ice winters, St Lawrence, north shore, and eastern shore ice will not be a hazard until mid-January. Warmer temperatures and fewer storms inhibit ice formation and drifting in the open Gulf, so that at its greatest extent ice is sporadic over the western area out to the Magdalen Islands and over the northern and eastern inshore zones. The Cabot Strait remains clear all winter and the central Gulf is easily navigable.

On 'the outside,' or the open Atlantic shelves, two regions may be identified: the northern and eastern coasts of Newfoundland, where ice is present for some part of the year, and the southern shores of Newfoundland and Nova Scotia which remain ice-free. In the former area, sea ice of local origin is joined by drifting floes and bergs from Arctic seas and glaciers. In a 'heavy ice' winter, the Labrador coast is fast with ice by early December and, by the end of January, Newfoundland bays as far south as Trinity Bay are ice-bound (Figure 1.8). Offshore, current-drifted floes and bergs will be seen by this time. In such 'heavy' winters it is mid-April or early May before east Newfoundland bays begin to break up, and isolated bergs trapped in them may persist until late June, while air temperatures have reached maxima of 60–65°F (15–18.3°C). Little ice is carried westward with the current that rounds the Avalon Peninsula to bring cold surface water to the south shores of Newfoundland and to Nova Scotia. Inshore fishing can continue there on a smaller scale through the winter from outports such as Ramea and Burgeo in Newfoundland.

It has become less important today to evaluate the existence of sea-ice for the seal fishery in Atlantic waters. The growing public outcry against the alleged cruelty of the activity since the mid-1960s, together with the overall decline of traditional maritime pursuits, has diminished the importance of a once-valuable industry. For the sealers, 'good' ice conditions mean that by early March the Gulf of St Lawrence and the northeast coast of Newfoundland are clogged with loose ice pans, not more than a foot or two thick, through which the newly arrived bulls and whelping cows can maintain access to and from the water. In order that vessels can time their arrival to coincide with the week or two in which newborn pups are being suckled on the ice, it is essential that high onshore winds do not close the open 'leads' or offshore winds drive the floes out of reach. Inshore sealing, once important to small, isolated settlements and still carried on on a small scale, depends in part upon land-fast sea ice with open leads, so that small groups can handle a boat to the whelping areas and efficiently kill quantities of seals close to the village.

SOILS[6]

The relative poverty of soils in the four Atlantic Provinces of Canada, together with a host of equally significant cultural and economic factors, has ensured that agriculture never held an important position in the regional economy. This might be one reason why soils have received little attention in geographical accounts, although soil surveys have been published for all of Nova Scotia and Prince Edward Island and most of New Brunswick.

Within the theme of fragmentation adopted in this volume, the mosaic of soil series[7] may be viewed as a quite direct influence on the spatial organization of the farming population. But this influence must not be seen as overriding others of an environmental and cultural nature. It is not helpful, for instance, to speak of a fragmented distribution of good and poor soils in an area populated by a group that has never been culturally oriented towards agriculture. In Cape Breton Island, for example, Dunn (1953) draws attention to the low opinion held of each other by hill-farmers of 'the rear' and fishermen of 'the shore,' respectively despising the others' occupation.

That 95 per cent or more of the soils in the Atlantic region are developed on generally stony glacial debris, are strongly leached, and are often thin and developed over a topography that ranges from hummocky to level, does little to distinguish them from other podzolic types occupying a broad belt along the southern edge of the Precambrian Shield from Quebec to Alberta. What is more distinctive is the intricacy of the mosaic of soil series developed over many textural variations of glacial deposits and occupying a more varied terrain than in other areas of the podzol zone. This intricacy is best exemplified along a transect across an upland block and adjoining lowlands where soil surveys typically identify a score of soil series, none of which occupies more than 10 per cent of the land area.

Another distinction can be made with respect to the remaining 5 per cent or so of the region's soils. These are developed on finer-textured morainic and riverine materials, and on sediments deposited during the marine submergence of the coastal fringe, either accompanying deglaciation, when they are merely sandy veneers over till, or during the last few

6   Sources consulted are listed in the references under Nova Scotia Soil Survey; New Brunswick Soil Survey; Whiteside 1950; Mollard and Munn 1955.

7   A soil series is identified by the recognition of similar soil profiles developed in similar parent materials under similar conditions of climate, vegetation, and landform.

thousand years, when they are often extensive estuarine muds. These soils occupy only small, restricted areas such as the Annapolis–Cornwallis lowland, and the northern shore of mainland Nova Scotia. However, their relative suitability for agriculture has heavily weighted their importance. Their distribution may be viewed as a strong influence on the fragmented ecumene of the Atlantic region, but such simplifications do no justice to historical factors which influenced the pattern and timing of land settlement.

In order to reduce the complexity of the soil series mosaic to a scale concordant with a regional treatment such as this, series have been grouped into assemblages common to larger physiographic divisions. The grouping is subjective and no doubt masks as well as falsifies similarities and differences between and within those divisions. A more meaningful grouping, particularly one integrating soils with other significant components of regional ecosystems, awaits future study.

Upland Zone of Northern New Brunswick

In the high, dissected plateaus where the headwater tributaries of the Tobique, the Nepisiguit, and the Miramichi rise, Palaeozoic granites, slates, argillites, and quartzites have each yielded a glacial till with significant differences for soil parent materials. On the granitic tills the Juniper series is a shallow, sandy loam with good to excessive drainage and abundant stones. The Glassville series, developed on a loam till derived from slates and argillites, has good drainage, but abundant erratic stones and a topographic setting that is rolling to hilly. The Homesville soil series, developed on quartzitic tills, is a gravelly, sandy loam with a moderate amount of stones, good drainage, and an undulating to hilly topography. These three series, which occupy 50–70 per cent of the land surface, share limitations of shallowness, stoniness, topography (and, thus, drainage and erosion problems), acidity, and low sustaining ability, all of which make them quite unsuited to agriculture. The middle and lower Tobique River and the upper Saint John are flanked by low plateaus mantled with a richer till, an ablation deposit, upon which is developed the Monquart series. These soils have a higher lime content and, because of a finer texture, have a greater inherent fertility and ability to sustain some crops. Some of this land is cleared and hay and potato crops are taken.

Middle Saint John River Valley and Southwestern New Brunswick

The central section of the Saint John valley, from near Andover to Fredericton, is incised some 200–300 feet (60–90 m) below an undulating plateau at 400–600 feet (120–180 m) above sea level, above which isolated hills rise to 700–800 feet (210–240 m). The plateau is covered with a thick mantle of glacial till, much of which is derived from shale and argillite bedrock of lower Palaeozoic and Carboniferous age. It forms the parent material for Caribou shaley loam and Carleton shaley clay loam soil series, which are the most extensive easily cultivable soils of the entire Atlantic region. They are the soils that support the potato crop of New Brunswick. The Caribou association forms a catena from sandy loam to clay loam over an undulating to rolling topography. Drainage is generally good and occasionally excessive; stoniness is not a hindrance to cultivation and, although quite acid, Caribou soils are easily maintained at high fertility because of their high base-exchange capacity. Soils of the Carleton series are associated with a more accidented topography but have fewer stones. A heavier texture makes them more retentive of moisture, but drainage is unimpeded, so that they are more strongly leached than Caribou soils. Both Caribou and Carleton soils are susceptible to strong sheet erosion on slopes, especially with row crops like potatoes, unless tillage methods are practised to prevent it. Both series have their imperfectly drained topographic associates – the gleisolic Washburn and Canterbury series – occupying depressional sites with a distinctive vegetation of spruce, fir, tamarack, and cedar, which has remained largely uncleared.

Small areas of stony till on these plateaus, where granitic or schistose bedrock is close to the surface, are characterized by soils of the Lomond series – a gravelly loam – and its poorly drained depressional associate, the Deed series. These soils are more widespread over the upland plateau north of the Fundy shore where crystalline rocks are close to the surface. The Lomond–Deed association is analogous to the Juniper series of northern New Brunswick. It is coarse, excessively or poorly drained, very stony, acidic, and of low natural fertility and, hence, of very low agricultural capability.

Betweeen the lower Saint John River and the Maine border, Caribou and Carleton soils are extensive but not as intensively used as those areas farther up the valley. They are interspersed with equally wide areas of much poorer Lomond–Deed soils and the equally poor Pinder–McAdam association on stony, sandy till of granitic derivation.

Bay of Fundy – Cape Breton Upland Zone

This zone includes the plateau of southeastern New Brunswick, the Cobequid Mountains and Antigonish uplands of northern and central Nova Scotia, the Cape Breton highlands, and the upland of North Mountain on the southern side of the Bay of Fundy. These uplands have common characteristics of igneous and metamorphic bedrock, generally coarser stony till parent material, and a topography consisting of discrete, gently sloping plateau surfaces at 700–1700 feet (210–600 m) into which short, steep-sided valleys are incised.

Over the uplands of southeastern New Brunswick, the Lomond–Deed soil association is extensively developed on a thin, bouldery till which mantles the undulating plateau surfaces at between 600 and 1200 feet (180–360 m). Similar soils in the Cobequid Mountains belong to the Cobequid, Wyvern, and Westbrook series. To the east, in the Antigonish and Cape Breton uplands, parent materials are variable, comprising granitic slates and quartzitic and conglomeratic tills. Soils of the Thom series are the most extensive here. Although limitations of depth, accidented topography, erosion susceptibility, and low base-exchange capacity combine to reduce their agricultural capability, locally they are excellent forest soils. Fully one-third of the area of Cape Breton Island falls within the mapping unit known as Rough Mountain Land and this lies entirely within the Cape Breton Highlands. There, the elevation at 1300–1700 feet (390–520 m), extensive level, poorly drained areas, occasional rock outcrops, or bouldery till parent material and cold, wet, climate combine to make these the poorest soils of the Maritime Provinces, comparable to those which are much more extensive over the barren plateaus of Newfoundland.

The long and narrow cuesta upland of North Mountain overlooks the Annapolis–Cornwallis lowland along a somewhat subdued escarpment, rising from the lowland margins at 300 feet (90 m) to a plateau at 600–800 feet (180–240 m). The basaltic bedrock of the north-sloping plateau bears a cover of bouldery and sandy loam till of local derivation which is typically thin and occasionally absent. Soils of the Rossway series are developed on it, presenting the same limitations to agriculture as other upland soils: shallowness, stoniness, and poor or excessive local drainage due to slope factors. They are not inherently infertile, however, as shown by the healthy stands of mixed forest developed over them.

The steep-sided valleys which dissect the margins of all these plateaus have the advantage of good soil drainage which is so often lacking over the summit surfaces. They are normally well forested and, if not disturbed, they comprise exceedingly healthy ecosystems. But, when cleared for farming or logged, the balance of slope–soil–vegetation variables is quickly upset.

The Atlantic Upland of Nova Scotia

The geological foundation of the upland extending from Yarmouth in the west to Canso in the east is simpler than that of the Fundy–Cape Breton

upland zone. Three rock types predominate: granites, slates, and quartzites (Figure 1.2). The gentle, uninterrupted southeastward slope from
extensive summit plateaus at 500–800 feet (150–240 m), and a situation
closer to the terminal position of the late-Wisconsin ice sheet, might account for the stronger glacial scouring suffered by these uplands compared with other uplands in the Atlantic region. This has left a till cover
that is generally thin and bouldery over the granite and quartzite outcrops,
but the weaker slates have yielded material which, mixed with the coarser
debris, has formed a sandy loam drumlinized till.

The granitic terrane is mantled with a coarse, bouldery ablation till,
upon which is developed the Gibraltar soil series and its poorly drained
associate, the Bayswater. In these uplands this association occupies about
25 per cent of the area. Rock outcrops are common and in Halifax
County, where granites are extensive, outcrops merit a separate land-type
category, covering 8 per cent of the area. Topography is undulating or
rolling, only rarely steep, and soils are consequently well or excessively
drained. The originally poor coniferous forest and scrub vegetation has
been debilitated by fires and a long history of lumbering. The land is primarily 'barrens' of the poorest agricultural and silvicultural capability.

The quartzite terrane is similarly mantled with a thin, rocky, sandy
loam till, upon which the Halifax series and its poorly drained Danesville
associate are developed. This land type occupies 20 per cent of the upland
counties and has a low suitability for agriculture. The hilly terrain on the
quartzite outcrops, however, provides better growth conditions for mixed
forest than the Gibraltar land-type.

The slate outcrops have yielded a loamy, bouldery till which is extensive over the outcrops themselves and to the southeast of them, where
drumlin fields in Lunenburg, Queens, and Yarmouth counties indicate a
short glacial transport of slate-derived till. The Bridgewater soil series is
the commonest here. It comprises two texturally distinct phases – the loam
phase of the drumlins and the sandy loam phase between drumlins and
the areas where these are not developed. The loam phase is usually stony
but this has not been a hindrance to cultivation of hay and certain grains
by industrious farmers of German stock. The sandy loam is shallower and
stonier and crystalline rock protrudes through it in places, so that drainage
is usually either poor or excessive.

Completing the mosaic of soil series on the Atlantic Upland are several
land types of sporadic distribution. Inland waters occupy 5 to 15 per cent
of these counties; peat is locally extensive, as around the margins of lakes
Rossignol and Kejimkujik; small patches of outwash gravels in upland
Annapolis county yield poor, droughty soils of the Nictaux series; shallow,

stony sand of the Aspotogan series occupies poorly drained depressions in association with the major land types.

Gulf of St Lawrence and Bay of Fundy Lowland Fringe

Apart from the broad lowland slope of eastern New Brunswick and northern Nova Scotia, continuing across Northumberland Strait in Prince Edward Island, the lowlands of the Atlantic Provinces are both small in area and physiographically separated from each other. They are developed over Carboniferous, Permian, and Triassic sedimentary rocks that are mechanically weak in comparison with the older, predominantly crystalline, rocks of bordering uplands. The rocks are mostly of clastic sedimentary origin, with occasional limestones in the Carboniferous formations. Glacial tills form the most extensive parent materials for soil development, but they occupy a lesser percentage of the lowlands than of the upland areas. As lowlands, many parts have a level topography and poor drainage has fostered the development of organic 'soils,' both small pockets of muck and more extensive lowland peat. River valleys are also wider and more gently sloping in the lowlands, and riverine sediments related to past and present stages of valley development are common and most important agriculturally. Lastly, over the coastal fringes of these lowlands marine sediments were deposited immediately following deglaciation. These materials range from coarse beach sands and gravels to finer silts and clays deposited in deeper waters.

Over the lowlands of eastern New Brunswick and northern and central Nova Scotia, the commonest and most extensive soil series is the Queens. It is developed on clay-loam tills derived from Carboniferous shales and sandstones. The topography is gently undulating, with broad interfluve areas between wide valleys, at elevations ranging up to 300 feet (90 m). Problems of drainage are common over level terrain and this is the most severe limitation to agriculture. In Prince Edward Island, 63 per cent of the land is occupied by Charlottetown (36%), Alberry (23%), and Culloden (14%) soil series. These have common elements of sandy loam texture, occasionally severe erosion, droughtiness, and acidity, but phases on gentle slopes support healthy forage crops and hay. O'Leary clay loam soils, derived from clay till, occupy 4.5 per cent of the island and are the only other first-class agricultural soils.

On the more accidented topography of the upland fringes glacial till, derived from a variety of sedimentary rock textures, forms the parent material of the Woodbourne and Millbrook sandy clay loam soils. The Woodbourne soils are more gravelly and, where topography permits, of good drainage, both series rating as good to fair land for hay and pasture. Where glacial tills reflect a derivation from predominantly sandstone bedrock, several soil series may be identified and distinguished on the basis of their stoniness and the associated topography. The commonest series is the Shulie, covering about 5 per cent of the counties from southeastern New Brunswick to eastern Cape Breton Island. It is developed on a sandy loam till occupying topography similar to that of Queens soils, except that slopes may locally be steeper and cause severe erosion problems where pastures have been overgrazed and woodland indiscriminately cleared. Other sandy loam soils include the Westbrook, Springhill, and Merigomish series, and the Harmony association. The agricultural capability of each is fair for cash crops when they are well managed (which is rarely), and good for hay and pasture land if potential erosion problems are mitigated.

Coarse-textured materials of glaciofluvial and glaciomarine origin are found in small areas sporadically distributed. In southern New Brunswick, the Gagetown series is of this type and its abnormally wide extent has repelled farmers from an otherwise highly suitable area. In the Annapolis lowland the Nictaux series is developed on outwash delta sediments in small areas where proglacial streams debouched from the uplands into the sea which flooded the lowland. Finer-textured marine sediments, ranging from sands to clays, have yielded only fair agricultural soils of the Canning, Cornwallis, and Lawrencetown series, with problems of droughtiness in sands or poor drainage in clays. The best soils in this lowland are the Berwick, Woodville, and Somerset series – sandy loams developed on till in well-drained sites on the footslopes of North and South Mountains.

In the area of southeastern New Brunswick and north-central Nova Scotia for which soil maps are available, soils of the Tormentine and Pugwash series are developed on water-worked tills of variable texture derived from Carboniferous sedimentary rocks. These are among the best agricultural soils of the Atlantic region, being stone-free, well-drained, and occupying a level or gently undulating topography. The sandy loam texture ensures a moderately high base-exchange capacity, so that these soils are capable of sustained yields of field crops as well as excellent hay.

The only other important soils associated with marine sediments in the region are those of the estuarine marshlands at the heads of Cobequid and Chignecto bays, and bays along the Northumberland Strait shore and at both ends of the Annapolis lowland. The parent materials are red silts and clays deposited during the recent submergence of the coast. The tidal range in the Bay of Fundy, 20 feet (6 m) in the west to 50 feet (15 m) at its head, has produced a great thickness of these sediments and exposes wide areas of mudflat at low tide. The estuarine marshlands have been diked and the consequent improvement of drainage has produced rich silts of the Acadia soil series, which rate highly as haylands.

Riverine deposits are, of course, highly localized along the modern river channels and lower slopes of lowland valleys. They occupy only a small area and yet the most attractive soils of the region are developed on them. Local limitations of flooding along channels, seepage in depressions on river terraces, and erosion on terrace bluffs exclude these soils from the highest agricultural capability class, but they are intensively used for vegetables, fruits, hay, and grains. In different parts of the region, alluvial soils have been given different series names. In southern New Brunswick they are known as Intervale (Interval) soils, a name derived from the local topographic name for the alluvial valley bottoms; in Nova Scotia variation in the source of alluvial sediment has given rise to the various series names – Cumberland, Stewiacke, and Bridgeville.

Island of Newfoundland
Since no soil survey has been made in Newfoundland, there is little more than a subjective basis for comparing soils there with those of the other Atlantic Provinces. Actually, over 95 per cent of the area no comparisons can be made because of differences in parent material, degree of rock exposure, vegetation cover, climate, and topography. Over the remaining small area, mostly sedimentary lowlands mantled with glacial till and related deposits, some comparison might be made with the lowland fringe of the Gulf and Bay of Fundy. But, without detailed surveys to support this comparison, it is preferable to treat the island's soils separately.

It is in Newfoundland that the most extreme limitations to agriculture are met. Of an area of roughly 43,000 square miles, about half is classified as bedrock, outcropping, or thinly and patchily mantled with stony, ablation till. One quarter is classified as ground moraine with variable topography, bedrock protrusions, and peaty depressions. This moraine is primarily bouldery and sandy ablation moraine and ice-contact debris, but locally it contains sufficient finer material to be of some agricultural use if limitations of poor drainage or steep slope are not felt. Slightly more than one-tenth is classified as 'end moraine,' where the surficial material is thick enough to mask the bedrock completely and the topography is highly accidented, with attendant slope and drainage hindrances to agriculture. Another tenth is underlain by organic terrain, composed of sphagnum peat in morainic depressions or blanket peat over many level and undulating areas. Blanket peat development commonly exceeds six feet in thickness over pervious sand and gravel terraces, raised beaches, and ablation till – even where there is a perceptible slope. Some of these bog areas offer potential as haylands and vegetable plots. Locally, as on the west side of the island, this has been developed by draining bogs. The remaining 5 per cent of the non-aqueous area of the island is classed as glaciofluvium, raised marine sediments, and recent alluvium. The latter covers only 0.2 per cent of the island and is generally coarse and often seasonally flooded. The other types are characterized by a level topography, which is a hindrance to agriculture if the sediments are thin over impermeable bedrock or till. Some raised sandy delta areas around St George's Bay are cleared for excellent vegetable and hay crops, as are similar but much smaller valley fills in isolated coves elsewhere on the island.

The tending of vegetable gardens and the cropping of hay has traditionally been important in the outport subsistence economy of Newfoundland, although, with increasing availability of tinned and packaged foods, this has declined. But the only areas where farming as opposed to fishing can be said to have ever loomed large in island terms are the Codroy valley, the St George's Bay shore, and the Deer Lake–lower Humber River lowland in western Newfoundland, where surficial materials are thick and topography is even enough, and in parts of the Avalon Peninsula, where proximity to the St John's urban market has been a greater spur to farmers than the meagre land resource.

FOREST TYPES[8]

The geobotanical history of the Atlantic Provinces is still imperfectly known. The present land area was completely covered by ice sheets at the last Wisconsin maximum, some 18,000 years ago, so that plant colonization and succession have taken place since that time from areas beyond the ice margins. The margin of the ice sheet stood some 30–50 miles (48–80 km) off the southern shore of Nova Scotia 18,000 years ago (King 1969). Prest (1969) shows all of the Gulf of St Lawrence ice-covered at that time, except for a small pocket around the Magdalen Islands. Ice margins off Newfoundland have not been traced, but sea-floor topography on the Grand Banks suggests an extension of the island ice sheet to the southeast at least as far as the Avalon and St Pierre channels, 45 miles offshore, and possibly twice that, on the Grand Bank itself.

Wide areas of the continental shelf south of Nova Scotia and Newfoundland therefore lay south of the ice sheet 18,000 years ago and a

8   Major sources consulted: Erskine 1961; Fernow 1912; Livingstone 1968; Loucks 1962; Ogden 1965; Roland and Smith 1969; Rowe 1959.

large portion of those are shallow enough to have stood above the sea at
that time, when sea level occupied a glacio-eustatic low level about 300–
400 feet below present. Howden *et al.* (1970), through their studies on
Sable Island, view such areas as likely refugia at the last glacial maximum.

Little is known of the biota that occupied those 'refugia.' Teeth of mam-
moth and mastodon have been recovered at many sites from 400 feet of
water on George's Bank in the southwest of this region (Whitmore *et al.*
1967), but evidence of climate and vegetation is lacking. A further problem is centred on the routes followed by plant and animal species to
present land areas. During the waning of ice sheets, the return of water
to the world ocean caused sea level to rise at 3–4 feet per century until
about 7000 years ago (Shepard 1963). Across a gently sloping shelf like
that off Nova Scotia and Newfoundland, a relatively small rise in sea level
would result in a relatively great landward advance of the shoreline, so
that sea level may well have stood against the retreating ice front during
deglaciation of the shelf. In southern New Brunswick, however, Gadd
(1970) suggests that deglaciation of the coast and construction of an end
moraine preceded marine submergence. Such a sequence would permit
vegetation to colonize at least a narrow coastal fringe in that area immediately before 13,300 years ago.

Analysis of pollen and plant macrofossils from bogs in the Atlantic
Provinces has so far not yielded a regionally coherent picture of postglacial vegetation change (see Livingstone 1968; Ogden 1965). The organic
zone overlying glacial deposits is usually interpreted as reflecting tundra-
like vegetation and this is succeeded by a more closed boreal forest assemblage of spruce and birch. A succeeding pine-dominated zone, like that
so common in New England pollen profiles, is often interpreted as reflecting a warmer episode, but lack of species distinction sometimes makes
this uncertain. Above the pine zone, pollen percentages reflect hardwood
dominance with hemlock, beech, and oak accompanied by minor fir, pine,
and spruce. Climatic interpretations of these changes have so far been
avoided because little is known of present-day climatic influences on individual species.

The Atlantic region is moist enough for continuous forest cover, but
variations in precipitation amounts, soil water budgets and thermal re-
gimes produce diversity of forest types. Wind is a further important environmental element in exposed coastal and upland locations. In the
absence of a classification of forest-climate regions, the following suffice
to bring out major contrasts: (i) cool, windy Atlantic and Fundy coasts
where the limiting stress is probably imposed by wind; (ii) more sheltered
coasts of the Gulf of St Lawrence where summer temperatures are warmer

than in (i); (iii) cool or cold, moist uplands where altitude and latitude depress summer temperatures and altitude induces increased precipitation; (iv) warm, drier, lowlands where an interior situation raises summer temperatures and lowers winter precipitation. These coarsely defined zones should be seen in relation to the map of forest regions (Figure 1.9).

The forests of the Atlantic Provinces fall within the following regions as defined by Rowe (1959): Boreal (predominantly forest), Boreal (forest and barren), Great Lakes–St Lawrence, and Acadian. The Boreal Forest Region, over the Notre Dame Mountains of Quebec and the whole of Newfoundland, is characterized by black spruce and white spruce, with tamarack (larch), balsam fir, and jack pine.[9] White birch is the commonest associate. The Great Lakes–St Lawrence Forest Region in the Atlantic Provinces is found on the lower uplands surrounding the Notre Dame Mountains and bordering the Acadian Forest Region on the north. It is a mixed forest, characterized by red and white pine (the red pine now rare in this area), eastern hemlock, and yellow birch. Hardwoods include sugar maple, red oak, basswood, and white elm.

The Acadian Forest Region, as its name implies, is restricted in Canada to the Maritime Provinces and is the most extensive region, excluding Newfoundland. Red spruce is characteristic, with balsam fir, yellow birch, sugar maple, and lesser proportions of red and white pine and hemlock. Beech was formerly important but has been reduced by beech bark disease. Black and white spruce are wide-ranging within the region, together with red oak, white elm, black ash, white birch, wire birch, and the poplars.

In this review, the work of Loucks (1962) has been extensively consulted with respect to the Maritime Provinces. This work did not include the island of Newfoundland, for which the widely known work of Rowe (1959) has been used here. The two classifications are not strictly comparable in their details but, since Loucks' scheme has only been applied here down to the 'ecoregion' level, omitting the many 'site districts' that he identified, some measure of congruence has been achieved. Loucks identified seven forest zones and eleven 'ecoregions'[10] within them. In Figure 1.9, two ecoregions in northwestern New Brunswick, part of his 'Fir-Pine-Birch Zone,' and one ecoregion in Cape Breton Island, comprising the whole of his 'Spruce Taiga Zone,' have been classified as 'forest

9 Linnaean names are listed at the end of the chapter.
10 Ecoregion: the geographic unit within which relationships between species and site are essentially similar, and within which silvicultural treatments may be expected to obtain comparable results; recognized by the characteristic species composition and development on the zonal site type. (Loucks, p. 166.)

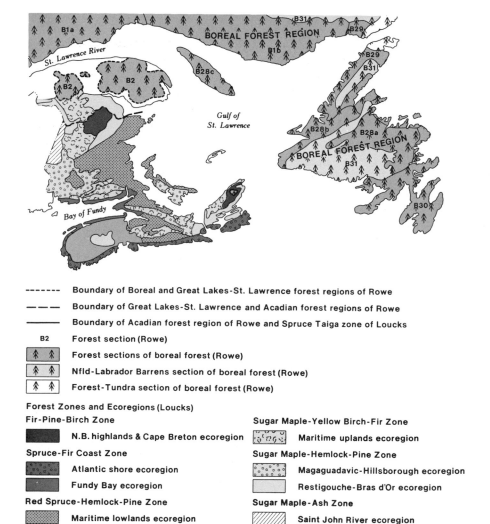

------- Boundary of Boreal and Great Lakes-St. Lawrence forest regions of Rowe

— — — Boundary of Great Lakes-St. Lawrence and Acadian forest regions of Rowe

——— Boundary of Acadian forest region of Rowe and Spruce Taiga zone of Loucks

B2   Forest section (Rowe)

Forest sections of boreal forest (Rowe)

Nfld-Labrador Barrens section of boreal forest (Rowe)

Forest-Tundra section of boreal forest (Rowe)

Forest Zones and Ecoregions (Loucks)

Fir-Pine-Birch Zone

N.B. highlands & Cape Breton ecoregion

Spruce-Fir Coast Zone

Atlantic shore ecoregion

Fundy Bay ecoregion

Red Spruce-Hemlock-Pine Zone

Maritime lowlands ecoregion

Clyde River-Halifax ecoregion

Sugar Maple-Yellow Birch-Fir Zone

Maritime uplands ecoregion

Sugar Maple-Hemlock-Pine Zone

Magaguadavic-Hillsborough ecoregion

Restigouche-Bras d'Or ecoregion

Sugar Maple-Ash Zone

Saint John River ecoregion

1.9

Forest Regions of the Atlantic Provinces

(Source: Compiled from Rowe, 1959 & Loucks, 1962)

sections' of Rowe's 'Boreal Forest Region.' In Table 1.3, Loucks' zone-ecoregion scheme and Rowe's region-section scheme have been combined, and characteristic tree species are shown in the order of their dominance in each forest unit.

When viewed as an all-important resource, the forests of the Atlantic Provinces face problems that are both severe and urgent. Although it was applied specifically to Nova Scotia forests, the following statement is valid for the area as a whole: '... lack of accurate, detailed knowledge about growth rates relative to species, soils, and locations is a serious deterrent to improving forest management ...' (Atlantic Development Board 1968, pp. 2–31). The same source concluded that by the year 2000 the supply of wood in Canada will be scarce and that the Atlantic Provinces will be the first to feel the consequences. Even where regulations have been prescribed for conservative use of forest resources, the study found that, at least in the case of Newfoundland, these regulations are 'incapable of enforcement' (*ibid.* pp. 2–37). While the principles of wise forest use may be known to the scientific community, the benefits and detriments of clear-cutting logging practices are still debated. Local pride may be the reason for the declamation: 'you underestimate the regenerative capabilities of Nova Scotia soil,' reported in the Halifax Chronicle-Herald as recently as January 1970 in defence of the practice. The effects of clear-cutting on soil erosion and stream sediment load, and the effects on fish populations have yet to receive detailed attention in this area.

For a sustained yield of merchantable timber it has been calculated that 50 per cent of the standing crop should be currently merchantable, 45 per cent should be young growth, and 5 per cent left unstocked. For the Atlantic Provinces as a whole, including Labrador, the ratio is 73 : 19 : 8. Newfoundland stands fairly well at 54 : 43 : 3, Nova Scotia quite unbalanced at 88 : 9 : 3, and Labrador alone at 85 : 4 : 11 (Atlantic Development Board 1968, pp. 1–8). Even if areas were brought closer to the desired ratio in the most unbalanced areas within the next five years, it would require 40 to 80 years for the benefits of proper rotation to be felt.

CONCLUSIONS

The foregoing treatment of the physical geography of the Atlantic Provinces is itself a summary of works on varied topics and would not be well served by any further summary. The main conclusion a reader might reach from it is that what is known of physical geography in the region is drawn from disparate, mainly non-geographical sources, and that a satisfying unified treatment of the subject awaits future study.

It may be timely to point to some areas in which more research is required. The most important, and most 'geographic' topic requiring attention is that of the structure and functioning of ecosystems in the broadest

Table 1.3 A classification of forest types for the Atlantic Provinces (after Rowe 1959 and Loucks 1962)

Forest zone or *Region*[1]	Ecoregion or *Section*	Characteristic species	Associated climate
Sugar Maple–Ash	St John River	sM, Be, wAs[2]	Warm, dry
		sM, Be, I, wAs, Bu, Ba	Warm, dry
Sugar Maple–Hemlock–Pine	Restigouche–Bras d'Or	sM, Be, wP, eH, yB	mod. warm, mod. dry
	Magaguadavic–Hillsborough	sM, Be, bF, yB, wP, wS	mod. cool, mod. dry
		sM, Be, wP, eH, bF, rS	mod. warm, mod. dry
Sugar Maple–Yellow Birch–Fir	Maritime Uplands	sM, yB, bF, Be	cool, moist
		sM, yB, bF, Be, wS, rS, rM	cool, moist
Red Spruce–Hemlock–Pine	Clyde River–Halifax	rS, bf, eH, wP, rM	mod. warm, mod. dry
	Maritime Lowlands	rS, wP, eH, rO, rM, bS, Be	warm, dry
		bF, rS, bS, eH, wP, rM, jP	mod. cool, mod. dry
		wS, bE	
Spruce–Fir Coast	Fundy Bay	wS, bF, wB	cool, wet
	Atlantic shore	rS, bF, wB, wS, bS, yB, Mo	cool, wet
		wS, bF, bS, wB	cool, wet
Fir–Pine–Birch	New Brunswick Highlands	bF, wB, wS, wP	cold, moist
		bF, wB, wP, tA, wS, jP, rS	cold, mod. dry
	Gaspé–Cape Breton	bF, wB, wS, bS	cold, wet
	(In N.W. New Brunswick, these mapped as part of Boreal Forest Region, section B.2, Rowe)		
Spruce Taiga	Cape Breton Plateau (mapped as Boreal Forest Region, but not by Rowe)	*bF, bS, wS, wB, Mo*	cold, wet

Table 1.3 (Concluded)

Forest zone or Region[1]	Ecoregion or Section	Characteristic species	Associated climate
Boreal Forest	Gaspe (B. 2)	bF, bS, wS, wB	cold, wet
	Grand Falls (B. 28a)	bF, bS, wB, wS, (wP)	cold, wet
	Corner Brook (B. 28b)	bF, bS, wS, wB, tA, Mm	cool, moist
	Northern Peninsula (B. 29)	bF, bS, wS, wB, tA	cold, moist
	Avalon (B. 29)	bF, bS, wS, wB, (wP)	cold, wet
	Newfoundland-Labrador Barrens (B. 31)	b̄S̄, b̄F̄	cold, wet

1 Roman type identifies Loucks (1962), Italic type identifies Rowe (1959), except climate.
2 Abbreviations are explained at the end of the chapter; ( ) signifies reduction of importance due to fire or lumbering; b̄S̄, b̄F̄ bar denotes stunted habit of trees.

sense. The treatment of soils and vegetation given here, for example, begs for improvement through studies of the interrelations among all elements – climate, landform, soil, plant cover – so that boundaries that have functional significance may be located. An important outcome of such studies would be an ability to predict the effects of various human modifications which will continue to press upon the physical and biotic resources of the Atlantic region.

Another largely unoccupied area encompasses the significance of land and sea resources to different groups of people throughout the history of the region. Little is known of the pre-European condition of these resources or of changes in ecosystem structure that have been effected throughout a long history of changing uses.

Further, studies of natural phenomena in the Atlantic Provinces have a history of their own which is mostly untold at present. With permanent European settlement dating back to the early seventeenth century, one would expect that a search for early accounts of the landscape would reveal much of changing views of, and attitudes towards, the 'stuff' of physical geography.

If these are significant times in which to be concerned about a fuller understanding of man's environment in the Atlantic Provinces, it is because the region is undergoing great changes in its human geography. The decline of the rural economy, localized industrial developments, internal population shifts, together with the rise of tourism and recreation and the omnipresent influence of governmental planning – all demand a closer acquaintance with ecologic elements than was gained from the more casual human adjustments of earlier times.

CODE FOR ABBREVIATIONS AND LINNAEAN NAMES

Softwoods
bS   Black spruce (*Picea mariana*)
rS   Red spruce (*P. rubens*)
wS   White spruce (*P. glauca*)
bF   Balsam fir (*Abies balsamea*)
wP   White pine (*Pinus strobus*)
jP   Jack pine (*P. banksiana*)
eH   Eastern hemlock (*Tsuga canadensis*)

Hardwoods
sM   Sugar maple (*Acer saccharum*)

rM   Red maple (*A. rubrum*)

Mm   Mountain maple (*A. spicatum*)

Be   Beech (*Fagus grandifolia*)

wAs   White ash (*Fraxinus americana*)

mO   Mountain ash (*Sorbus decora*)

I   Ironwood (*Ostrya virginiana*)

Bu   Butternut (*Juglans cinerea*)

Ba   Basswood (*Tilia americana*)

yB   Yellow birch (*Betula lutea*)

wB   White birch (*B. papyrifera*)

rO   Red oak (*Quercus rubra*)

tA   Trembling aspen (*Populus tremuloides*)

ACKNOWLEDGMENT

I should like to thank David Erskine, York University, for comments which helped me avoid many errors of omission and commission. Those remaining are my responsibility.

2 People in Transition: The Broken Mosaic

ALAN G. MACPHERSON

Within the framework of the physical geography of the Atlantic Provinces described in chapter 1 live some two million people. In 1966 approximately 780,000 of this total lived in an urban system consisting of places with populations of over ten thousand, as described in chapter 4. It is the purpose of chapter 2 to consider that part of the population which can be described as 'small urban' (places with less than ten thousand people) and rural, and which in a historical sense certainly, and perhaps in a purely demographic sense, can be accepted as the fundamental population component in the cultural and political geography of the Atlantic Provinces. In 1966 it consisted of some 1.2 million persons or about 60 per cent of the total population, while the same categories at the national scale formed only 38 per cent of the population of Canada. These facts seem to indicate that the population of the Atlantic Provinces is the least concentrated in large urban agglomerations of any region within the national territory.

In Table 2.1 the 'small urban'/rural population of the Atlantic Provinces is broken down by settlement size, following the census categories. The table shows that the population living in centres between 2500 and 9999 persons and between 1000 and 2499 persons forms only a relatively small part of the total provincial and regional populations, as compared with the 'large urban' (over 10,000) and rural (below 1000) elements. The two larger provinces, Nova Scotia and New Brunswick, show percentages for both 'small urban' categories fairly close to the percentage for the whole region. Prince Edward Island has virtually no concentrations in the two 'small urban' categories, being polarized between the 'large urban' centres of Charlottetown and Summerside on the one hand and the rural component on the other. In terms of settlement heirarchy and function this would seem to reflect the small size of the Island Province in both area and total population, and the sharp dichotomy between the farming and urban landscapes. Newfoundland, on the other hand, has larger components of its total population in the two 'small urban' categories than

Table 2.1 Population by settlement size in places below 10,000: 1966 (Census 1966: Table 10)

	Newfoundland		Prince Edward Island		Nova Scotia		New Brunswick		Atlantic Provinces	
	Nos.	%	Nos.	%	Nos.	%	Nos.	%	Nos.	%
Total population	493,396		108,535		756,039		616,788		1,974,758	
2500–9999	75,098	15.2	—		44,099	5.8	49,298	8.0	168,495	8.5
Below 2500	275,908	55.9	80,066	73.8	356,416	47.1	316,162	51.3	1,028,552	52.1
1000–2499	57,661	11.7	3,754	3.5	37,144	4.9	40,292	6.5	138,851	7.0
Below 1000	218,247	44.3	76,312	70.3	319,272	42.2	275,870	44.7	889,701	45.1

any of the other Atlantic Provinces, indicating that its population was already more thoroughly centralized in places of these size categories before the joint federal/provincial resettlement programs initiated in 1965 had taken effect. Its entire population, in fact, is virtually agglomerated throughout the hierarchy, from St John's to the tiniest outport, a reflection of its traditional dependence upon fishing rather than farming, and upon sea rather than upon land communications. Its integrated road system is post-Confederation (1949) in origin.

The percentages of the provincial and regional populations in settlements below 2500 and 1000 persons respectively all show the reciprocal situation: that the Atlantic Provinces, separately and collectively, have surprisingly large components in the 'very small urban' and rural categories, Prince Edward Island showing almost three-quarters of its total population and the others between 47 and 56 per cent in these categories. These facts, however else they may be interpreted, would seem to point to something fairly fundamental in the total circumstances of this part of Canada, something which may well explain the relative lack of success by government planners in their attempts to alter the underlying structure of society in the Atlantic Provinces, and something which may require a radically different approach and philosophy on the part of agencies which wish to help the region.

THE SMALL URBAN CENTRES

Those centres with populations between 2500 and 10,000 are listed in Table 2.2 with their 1966 populations.[1] Prince Edward Island has no centres in this category and does not appear in the table. In this range of the settlement heirarchy places exist because they perform a special function or are located in relation to a particular resource. Thus Dalhousie, Newcastle, and Liverpool are pulp and paper milltowns; Yarmouth, Carbonear, Lunenburg, Grand Bank, Caraquet, Harbour Grace, and Shelburne are bases for the dragger fleets and have fish processing plants; and Newcastle, Labrador City, Springdale, and Wabush are mining towns. Bay Roberts is a wholesale distributing centre, while Yarmouth, Gander, Port-aux-Basques, Deer Lake, and Botwood all represent points where

1   The Preliminary Bulletins of the 1971 Census show a situation little different from the one represented here. Many of the towns have grown or declined slightly in population; none have grown or fallen out of the size category. Wabana has continued its decline, to 5256. The category has been joined by Summerside, fallen to 9315, and by Sherwood grown to 3792, both in Prince Edward Island; by Port Hawkesbury, 3427, in Nova Scotia, and by Marystown, 4982, St Anthony, 2586, and Burin, 2580, in Newfoundland, all growth centres.

Table 2.2 Total urban population in centres
between 2500 and 10,000: 1966

Newfoundland		Nova Scotia		New Brunswick	
Wabana	7884	Yarmouth	8319	Chatham	8136
Gander	7183	Springhill	5380	Dalhousie	6107
Stephenville	5910	Kentville	5176	Newcastle	5911
Port-aux-Basques	5692	Antigonish	4856	Woodstock	4442
Labrador City	4992	Bridgewater	4755	Grand Falls	4158
Carbonear	4584	Windsor	3765	Dieppe	3847
Deer Lake	4289	Liverpool	3607	Sussex	3607
Botwood	4277	Lunenburg	3154	Marysville	3572
Happy Valley	4215	Shelburne	2654	St Stephen	3285
Bonavista	4192	Wolfville	2533	Sackville	3186
Bishop Falls	4127			Caraquet	3047
Bay Roberts	3455				
Grand Bank	3143				
Lewisporte	2892				
Harbour Grace	2811				
Springdale	2773				
Wabush	2669				
Totals	67,437		44,199		49,298

air or water transportation meets road and rail, with many of the ancillary services. Gander and Deer Lake are also successful and growing service centres for fairly extensive areas of Newfoundland: Gander for eastern Notre Dame Bay, including Twillingate and Fogo Island, and northern Bonavista Bay; Deer Lake for the whole Northern Peninsula. They are joined by such agricultural service centres as Kentville and Windsor in Nova Scotia. Bridgewater is a major logging centre.

A number of places in this size category represent relict roles in the regional economy in that their function has been either lost or severely reduced. Thus Wabana (Bell Island, Conception Bay) and Springhill are essentially relict mining towns, whereas Bonavista is a relict fishing base, formerly involved in the Labrador floater fishery, but now largely restricted to inshore operations. In a special category are Gander, Stephenville, and Happy Valley, all of which received their initial stimulus from their associations with military bases during and after World War II. Stephenville looks towards a future as a diversified industrial centre, largely geared to the pulp and paper industry, since the closure of the Harnum Field base in 1966. Happy Valley, still tied to the remnant of the Goose Bay base at the head of Lake Melville, Labrador, is developing into the logging centre for the exploitation of the Lake Melville forests. It is also the relocated headquarters for the Moravian Church in Northern

Labrador. The small university towns of Antigonish, Sackville, and Wolf-ville are also in a special category.

There are thirty-eight urban centres in this size category, three of which are located in Labrador–Newfoundland. All but the two company mining towns in Labrador are incorporated municipalities. Twenty-six of them had populations between 2500 and 5000 in 1966, that is they were in the lower half of their size category. Newfoundland's greater degree of centralization is indicated by the fact that seventeen of the places listed in Table 2.2 are located in that province, despite the fact that it has the smallest total population of the three.

THE RURAL COMPONENT

Table 2.3 describes the rural component in the population of the Atlantic Provinces, living in centres of less than one thousand persons down to the individual farmhouse and isolated country dwelling. The minor differences in the provincial and regional percentages for rural population as compared with those given in Table 2.1 are due to differences in definition of communities as between Tables 10 and 13 of the Census of Canada, and do not materially change the general points inferred from Table 2.1.

The total rural population in all provinces constituted at least 40 to 50 per cent, and in Prince Edward Island's case 63.4 per cent, of the total population in 1966. The regional percentage (46.5) was almost double the national figure of 26.4. Moreover, although decline in relative strength since 1951 was common to the country, the region, and the individual Atlantic Provinces, the percentage decline in the contribution of the rural sector in the Atlantic Provinces only slightly exceeded the national decline (11 points) in the cases of Newfoundland and Prince Edward Island; New Brunswick fell rather less (8 points) than the country as a whole; and Nova Scotia's relative decline was less than three percentage points. During the period 1951–66 it would appear that the forces of urbanization were acting less efficiently upon the rural sector of the population in the Atlantic Provinces, particularly in Nova Scotia, than in Canada as a whole. The rural tradition, as expressed in current structures of settlement hierarchy, economy, and social structure, would appear to have retained much of its ability to resist change, presumably by a retention or reinforcement of conservative values in the way of life.

Consonant with this are the facts that while Canada's total population grew by 43 per cent between 1951 and 1966, the population of the Atlantic Provinces only advanced by 22 per cent; the rural sector of the Atlantic Provinces, on the other hand, advanced by only 6 per cent in

Table 2.3 Population of the Atlantic Provinces living in places of less than 1000 persons: 1951, 1966
(Table 13: Census 1961, 1966)

		1951				1966			
		Provincial total	Total rural	Farm	Non-farm	Provincial total	Total rural	Farm	Non-farm
Nfld.	Nos.	361,416	207,057	15,509	191,548	493,396	226,707	8,455	218,252
	%		57.3	4.3	53.0		45.9	1.7	44.2
P.E.I.	Nos.	98,429	73,744	46,757	26,987	108,535	68,788	30,841	37,947
	%		74.9	47.5	27.4		63.4	28.4	35.0
N.S.	Nos.	642,584	287,236	110,198	177,038	756,039	317,132	45,251	271,881
	%		44.7	17.2	27.5		42.0	6.0	36.0
N.B.	Nos.	515,697	296,228	144,257	151,971	616,785	304,563	51,504	253,059
	%		57.4	28.0	29.4		49.4	8.4	41.0
Atlantic Provinces	Nos.	1,618,126	864,265	316,721	547,544	1,974,755	917,190	136,051	781,139
	%		53.4	19.6	33.8		46.5	6.9	39.6
Canada	Nos.	14,009,429	5,191,792	2,769,286	2,422,506	20,014,880	5,288,121	1,913,714	3,374,407
	%		37.1	19.8	17.3		26.4	9.8	16.8

Table 2.4   The net contribution of rural growth to provincial
growth in population, 1951–66, in percentages

	Nfld.	P.E.I.	N.S.	N.B.
Provincial growth	39.3	10.2	17.6	19.6
Provincial rural growth	9.5	6.7	10.3	2.8
Contribution of rural to provincial growth	14.9	−49.0	26.3	8.2

those fifteen years, and accounted for only 15 per cent of the total regional growth. The importance of the rural sector in the population structure of the Atlantic Provinces during the period 1951–66 can be seen most dramatically when it is considered with the rural sector of Canada as a whole. Taking only the crude regional and national figures as given in Table 2.3, the rural sector of the Atlantic Provinces (as categorized in 1966) contributed no less than 54.9 per cent of the growth in that category in the country as a whole. The comparable figures for each of the Atlantic Provinces, taken separately, are given in Table 2.4, derived from Table 2.3.

The growth of the rural sector was in every case substantially less than the total provincial growth, in Prince Edward Island's case suffering an actual decline. Its contribution to provincial growth was consequently small, especially in New Brunswick, and in Prince Edward Island where it actually had a depressing effect on growth. In Newfoundland rural growth was the same as for the Atlantic Provinces taken collectively, and only in Nova Scotia did it make a substantial contribution. In fact, despite relative decline as a component of the total population, the rural sector grew numerically in the Atlantic Provinces by a crude total of 52,925 persons, mostly in Nova Scotia (29,896) and Newfoundland (19,650).

Calculations of the contribution of the rural sector to provincial growth do not include that part of the rural population of 1951 which migrated to the urban sector in the ensuing fifteen years. Nor do they take into account inclusion of rural residents in urban centres by boundary changes and simple changes in size category by growth *in situ* to exceed a community of one thousand souls. This last factor applied peculiarly to Newfoundland between 1951 and 1966, insofar as that province was blessed with many outport communities hovering near the arbitrary figure; it was virtually non-existent in the Maritimes. Some twenty-five Newfoundland communities were affected, only one of which declined below the 1000 mark; some 16,158 persons who were categorized as rural residents in 1951 were excluded from this category in 1966 although they had neither moved nor been included in the urban sector by boundary change; in 1966 about 35,908 persons were living in the communities which had changed size category in the Census. In fact, it was only the communities classified as rural in 1966 which were responsible for the fifteen-year growth of the rural sector in the province. If change in size category were ignored, then the rural sector of 1951 actually increased *in situ* by approximately 27 per cent, and actually contributed 30 per cent, exactly double the figure derived from the crude statistics, to the total growth in Newfoundland's population. Its total contribution, including the effects of migration and boundary changes, would undoubtedly raise that figure considerably. Although the total role of the rural sector in development (in the purely demographic sense) remains largely unexplored and ignored by the planners, it is clear that it is an important and integral part of the growth process.

Farm and Non-Farm

The rural component in the population consists of two sub-categories, 'farm' residents and 'non-farm' residents (Table 2.3). In 1951 these subcategories were virtually equal in New Brunswick. In Nova Scotia the 'farm' sector was smaller, and in Newfoundland substantially smaller, than the 'non-farm.' Only in Prince Edward Island was the situation reversed, with the 'farm' sector dominant in the countryside. Newfoundland has only a very small agricultural sector in its economy; Prince Edward Island is almost wholly dependent upon farming.

By 1966 considerable reductions in the 'farm' population had occurred in all four provinces, and especially in Nova Scotia and New Brunswick. The relative position of the 'farm' population within each provincial population structure had also slipped, even in Newfoundland where it was already low fifteen years earlier. Prince Edward Island, although still endowed with a markedly stronger farm group than elsewhere in the region, now conformed to the general pattern in that its 'non-farm' group was now dominant. Indeed, in the Atlantic Provinces as a whole the relative position of the 'farm' population slipped from 19.6 to 6.9 per cent between 1951 and 1966, whereas in Canada as a whole it only slipped from the same position to 9.6 per cent. Whereas in 1951 one in five Canadians was classified as a farm resident, and the ratio in the Atlantic Provinces was the same, in 1966 the ratios were one in ten across Canada and only one in fifteen in the Atlantic Provinces.

In 1951 the 'non-farm' residents formed a group that was twice as important in the region as it was in Canada as a whole. Whereas in Canada as a whole this component has slipped in relative strength, in the Atlantic Provinces it has grown numerically everywhere, and has grown relatively in all but Newfoundland, where it nevertheless still forms a larger proportion of the total provincial population than elsewhere.

Changes in the 'farm' and 'non-farm' components in the population, however, are obscured by changes in the census definition of a farm resident. In 1951 such a person resided on an agricultural holding of three or more acres with agricultural production of $250 or more. In 1966, on the other hand, he resided on an agricultural holding of one acre or more with sales of agricultural products worth $50 or more. The reduction in numbers and relative position of 'farm' residents between 1951 and 1966, despite the wider definition in the latter year, is an indication that the Atlantic Provinces were affected rather more than the country as a whole by the flight from the land as a way of life. This process, however, was markedly insufficient to offset the growth of the 'non-farm' sector. If the extreme assumption be made that the flight from the land did not, in fact, involve actual migration into the urban centres, but was simply an abandonment of farming while retaining the farmhouse as a family dwelling, the 'non-farm' component was still able to sustain its own growth: by at least 29,896 persons in Nova Scotia, 19,650 persons in Newfoundland, and 8335 persons in New Brunswick. Only in Prince Edward Island did the decrease in 'farm' residents exceed the increase in 'non-farm' rural residents, sending a minimum of 4956 individuals into the provincial urban sector.

It would appear therefore that population growth was characteristic of all size categories of settlements with the exception of the farm sector. The population involved in the farming way of life in the Atlantic Provinces is suffering steady fragmentation and disintegration, with little evidence that the process has as yet exhausted itself or is approaching a new equilibrium.

AGE GROUPS AND COHORTS

It would be wrong, however, to conclude from the foregoing that the population of the Atlantic Provinces, and particularly the rural population, is in a healthy state of growth. Table 2.5 shows the numerical and percentage losses of each 5-year cohort present in the Atlantic Provinces in 1961 between that year and 1966. In the three Maritime provinces, somewhere between 70 and 80 per cent of the new 5-year cohort, aged 0–4 years, was required to make up for losses from deaths and out-migration in every other cohort in the population.

Newfoundland, in this respect, was somewhat better off than its sister provinces, for only 45 per cent of the new age cohort was required to compensate for losses among older people. But even there a halving of the birth rate would virtually bring population growth to an end. In the Maritimes a 20 to 30 per cent decrease would bring population to a no-growth

Table 2.5 Age cohort losses, Atlantic Provinces, 1961–66 (Census, 1966: Table 19)

| | Age groups 1966: | | | | | | | | | | | | |
	5–9	10–14	15–19	20–24	25–29	30–34	35–39	40–44	45–69	70+			
Newfoundland													
Cohorts 1961	67695	64404	59464	43829	30238	26719	25571	24828	88215	26895	0–4:1966	68545	
Losses 1961–66	688	873	5157	7853	2307	1351	295	1077	5422	7984	Total loss	30717	
Percentage loss	1.0	1.4	8.7	17.9	7.6	5.0	1.2	4.3	6.1	29.7	Ratio	44.8%	
Prince Edward Island													
Cohorts 1961	13221	12216	12264	8875	6344	5685	5364	5727	24003	10930	0–4:1966	12587	
Losses 1961–66	198	193	1203	2094	629	144	–10	189	1205	2836	Total loss	8681	
Percentage loss	1.5	1.6	9.8	23.6	9.9	2.5	nil	3.3	5.0	25.9	Ratio	69.0%	
Nova Scotia													
Cohorts 1961	91239	84760	80329	64239	49311	43956	43360	45041	171355	63417	0–4:1966	85521	
Losses 1961–66	3806	3160	6187	11641	6296	2599	1751	2532	11837	17780	Total loss	67589	
Percentage loss	4.2	3.7	7.7	18.1	12.8	4.8	4.0	5.6	6.9	28.0	Ratio	79.0%	
New Brunswick													
Cohorts 1961	78560	75882	72745	53514	37419	33621	33856	35983	129439	46917	0–4:1966	72859	
Losses 1961–66	2265	2974	7178	11183	4178	1664	1095	2047	8209	13222	Total loss	54007	
Percentage loss	2.9	3.9	9.9	20.9	11.2	4.9	3.2	5.7	6.3	28.2	Ratio	74.1%	
Atlantic Provinces													
Cohorts 1961	250715	237262	224802	170457	123312	109981	107056	111579	491538	148159			
Losses 1961–66	6957	7200	19725	32771	13410	5758	2561	5845	26665	41822			
Percentage loss	2.8	3.0	8.8	19.2	10.9	5.2	2.4	5.2	5.4	28.2			

position. Only if reduced growth in a consumer demand economy (fewer babies) was accompanied by a diversion of resources into production of goods and services (retention of a bigger labour force) would the region be saved from absolute decline. The two variables to watch in future are the provincial birth rates and out-migration rates.

Table 2.5 shows that the cohort suffering least loss in all provinces was aged 35–39 in 1966. Percentage losses increased somewhat in the older cohorts, but did not exceed the percentage losses of the 20–24 year olds till the retired group (70+) is reached. It can be crudely assumed that attrition by death was an increasingly important factor above the age of 40, while out-migration was the dominant factor below the age of 35. In every province the three cohorts most affected consisted of persons aged 15–29, the younger part of the potential labour force. Most of these were undoubtedly single persons, but the losses sustained by the children of school age in 1966 during the previous five years are an indication that families were also involved in the out-migration. The percentages indicate that Newfoundland and Prince Edward Island suffered relatively less from loss of families than did the two mainland provinces, a point that might also be confirmed by the lower percentages of losses in the parental cohort, aged 35–39 in 1966, in both island provinces. In any case the Atlantic Provinces sustained severe losses in those age cohorts that might have been expected to contribute to regional economic growth.

If it is recalled that the rural component in the population of the Atlantic Provinces constituted over 45 per cent of the total population in 1966, it can be assumed that provincial out-migration and intraprovincial urban drift are likely to have exerted a distorting effect on rural age-structure. This is demonstrated for the whole region in Figure 2.1. The distortion is evident between the 15–19 and 50–55 age groups, indicating that the process has been going on for some time. But the partially related effect of a fall in the crude rural birth rate, as indicated by comparing the two youngest age groups, is a new and potentially ominous sign.

The rural population of the Atlantic Provinces, then, in its present state constitutes a paradox: an important component in the total regional population, contributing vigorously to provincial, regional, and national growth while sustaining its own growth, and at the same time, consisting of broken age cohorts which threaten its economic, social, and demographic viability in a world devoted to growth. To understand the reality behind the statistics it is necessary to examine some of the underlying structure and the more intimate processes at work.

Fragmentation characterizes every pattern in the human and cultural geography of the Atlantic Provinces, and can be accepted as a funda-

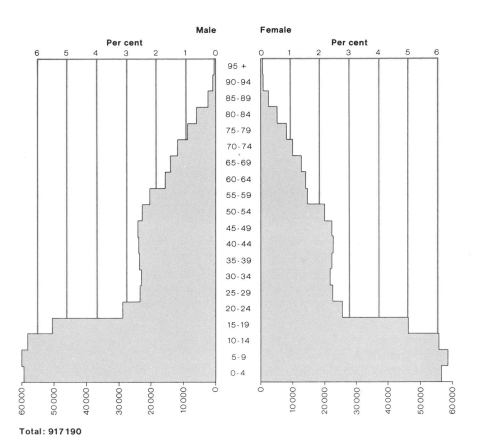

Male   **Female**

2.1

Age Structure of the Rural Population, Atlantic Provinces 1966

(Source: Census of Canada 1966: Table 20)

mental condition of the entire region. It is present in the political disunity imposed by federal structure; in the ethnic mosaic and its related factors of language and religious affiliation; in the discontinuous distribution of the small urban and rural population which reaches its extreme expression in the Newfoundland outport and its more isolated counterpart on the Labrador Coast. It appears in the spatial structure of the individual outport, and in the seasonality of men's work which anthropologists relate to occupational pluralism and economists to unemployment, and which, as a consequence, provides an income derived from many sources, and not always in cash. So far it is but the obverse of the coin where men integrate their affairs, individually and collectively, in order to live practical and reasonable lives. It is undoubtedly the genius of the people of the Maritimes and Newfoundland that they have succeeded in doing this. But when it appears in the sundering of the generations: in the rupture of the local folk culture, and in the anomie of the older generation as inheritance customs falter and continuity fails, then it can be seen in its socially disintegrative guise.

POLITICAL STRUCTURE AND BEHAVIOUR

The region has inherited a disunited political structure as a result of the manner in which the several British colonies and dependencies entered Confederation between 1867 and 1949. The fragmentation of political control over the primary resource base discussed in chapter 3 has resulted in provincial rivalries where single-minded purpose might otherwise have been looked for. The monopolizing tendencies of farmers and fishermen in matters of marketing and, in the case of common resources such as fish, actual exploitation, are matched by a total lack of co-operation in forest resource management. The marketing of Nova Scotian eggs and fresh milk in Newfoundland elicits the same political response as the threat which Nova Scotian herring-seiners and scallop-draggers are deemed to present in the management and exploitation of Newfoundland's inshore resources. While political consultation and executive governmental co-operation now seem more acceptable, albeit often at the urging of federal authorities, Newfoundland still tends to stand aloof; the legislative union of the four provinces which was looked for at Charlottetown and Quebec in 1864, and which has been advocated as an essential element in the reform of the federal structure of Canada during the 1970s, is still far from a political possibility.

At a lower scale sectional interests appear in the political aspirations and behaviour of each of the provinces. The threatened alienation of Labrador from Newfoundland, expressed in party formation and electoral behaviour, is matched by the perennial re-emergence of the ideal of the *Canadien* Madawaska Republic in northwestern New Brunswick. The distinctive political roles of the Catholic areas in the Avalon Peninsula and in the Port-au-Port/St George's Bay/Codroy Valley districts of western Newfoundland are matched by the *Acadien* as a force in northeastern New Brunswick. At the lowest scale political success is still measured largely in terms of material advantage to the individual constituency, whether it be provincial or federal, and small vulnerable communities fight bitter feuds over the siting of a new regional high school, a new federal wharf or community stage, a new cottage hospital or provincial park, or the construction of a new section of road. Incorporation for purposes of local government has been accepted reluctantly, at least in Newfoundland: of the 168 areas (cities, towns, rural districts, local improvement districts, and local government communities) incorporated by 1968 and representing some 60 per cent of the provincial population, no less than 101 were unincorporated before 1960 and only 18 were incorporated by 1950. The latter represented only 22 per cent of the population of the new province in 1951.

ETHNICITY AND CULTURAL SURVIVAL

The Atlantic Provinces have long been celebrated by scholars and politicians as containing the oldest pieces in Canada's ethnic mosaic. Few geographers have attempted to elucidate the regional pattern (Clark; Gentilcore; Mannion; Delaney and Macpherson), yet the cultural survivals and traditions which it indicates embrace a large part of the conservative values resisting or retarding social and economic change in the Atlantic Provinces today. The pattern is virtually identical with the rural population of the region. It is not surprising therefore that it is this vital and vulnerable component in the population which has received the greatest attention from governmental planning agencies. At the same time it has received the least understanding, as evinced by the assertion of one federal mandarin that the redevelopment and rehabilitation of Cape Breton Island's economy requires the demise of the island's Scottish culture.

Table 2.6 presents a statistical view of the rural population of the Atlantic Provinces for 1961, the last census year for which ethnic and linguistic information has been published. It indicates that the rural population of the Atlantic Provinces, individually and collectively, is almost entirely made up of people of English, French, Scots, and Irish extraction: around 95 per cent in Newfoundland, Prince Edward Island, and New Brunswick, about 80 per cent in Nova Scotia. Almost all other ethnic origins found in Canada (European and Asian) are present, but for the most part in insignificant numbers: in Nova Scotia in particular 8.6 per cent of the rural population (29,035) claimed German origin, 4.5 per cent (15,127) Dutch, and 1.6 per cent (5,316) Negro. The relative strength of the rural English and Irish is everywhere greater than in rural Canada as a whole, and this is also true for the Scots except in Newfoundland. On the other hand the percentage of rural people of French extraction only exceeds the national figure in New Brunswick, where they bear

Table 2.6 Ethnic composition of the rural population, Atlantic Provinces, 1961: numbers and percentages (Census 1961, Table 36)

	Total	English	Irish	Scots	French	Eskimo/Indian
Nfld.	225,833	174,586 77.3	34,032 15.0	3,431 1.5	8,614 3.8	1,172 0.5
P.E.I.	70,720	19,857 28.0	13,201 18.7	23,005 32.5	11,918 16.8	224 0.3
N.S.	336,495	104,758 31.1	36,958 11.0	78,832 23.4	48,287 14.3	2,818 0.8
N.B.	319,923	76,557 23.9	39,318 12.3	40,813 12.8	144,353 45.1	2,791 0.9
Atl. Provs.	952,971	357,758 39.4	123,507 13.0	146,081 15.3	213,172 22.4	7,005 0.7
Canada	5,537,857	1,177,841 21.3	544,956 9.8	544,668 9.8	1,761,765 31.8	191,739 3.5
Atl. Provs.: Can.		31.9	22.7	26.8	12.1	3.7

Table 2.7 Mother tongue of the rural population, Atlantic Provinces, 1961: numbers and percentages (Census 1961: Table 65)

	Total	English	French	Gaelic	Indian/Eskimo
Nfld.	225,833	222,455	1,699	23	1,080
		98.5	0.75	0.01	0.48
P.E.I.	70,720	63,433	6,467	41	109
		89.7	9.1	0.06	0.15
N.S.	336,495	299,615	28,608	2,518	2,077
		89.0	8.5	0.75	0.62
N.B.	319,923	178,290	136,845	21	2,537
		55.7	42.8		0.8
Atl. Provs.	952,971	763,793	173,619	2,603	5,803
		80.1	18.2	0.3	0.6
Canada	5,537,857	3,142,340	1,622,204	3,433	155,100
		56.7	29.3	0.06	2.8
Atl. Provs.: Can.		24.3	10.7	75.8	3.7

the strongest rural tradition: their relative weakness in Newfoundland and Nova Scotia reduced the regional figure below the national one.

Comparison of the percentages in Table 2.6 and Table 2.7 for French extraction and French mother tongue indicate, further, that the oral and literary tradition has been seriously weakened in Prince Edward Island and Nova Scotia. It has been virtually abandoned in Newfoundland, even if allowance is made for the fact that the census seriously underenumerated native French-speakers in the Port-au-Port Peninsula. Only in New Brunswick does the proportion of the rural population speaking French approach the proportion claiming French ethnic origin. Gaelic, once widely spoken by Highland Scots settlers, is now quite insignificant – albeit also probably underenumerated – although it survives in all four provinces. Assuming that all recent immigrants go to the cities, rural Nova Scotia, by these figures, has 75 per cent of all native-Canadian Gaelic-speakers.

The last line of Table 2.6 indicates that substantial segments of each of the older pioneer traditions in Canada are located in the Atlantic Provinces. They are represented on official index maps to legal surveys of Crown Charters, where surname evidence often shows intricate patterns of juxtaposition: as between *Acadien* French and the others throughout New Brunswick; as between Highland Scots and *Acadien* French in parts of Nova Scotia; as between Irish and English in Newfoundland. They also express themselves intricately on the landscape, as Mannion has demonstrated for the Irish in the Miramichi in New Brunswick and in parts of the Avalon Peninsula of Newfoundland. In terms of non-material culture, on the other hand, assimilation of early English into an Irish tradition has occurred along the Avalon's Southern Shore, south of St John's, while persistent intermarriage between Highland Scots, *Acadien* French, and Irish – all originally from Cape Breton – has failed to eradicate the distinctiveness of these traditions in the Codroy Valley and St George's Bay, Newfoundland (Delaney). Micro-studies of cultural traditions among the Scots and Irish in Nova Scotia and Newfoundland indicate, further, that distinctions exist within each of these groups, invalidating earlier generalizations. Besides distinctions between Highland and Lowland Scots, important in Newfoundland where the latter are concentrated in the St John's/Conception Bay area, and between Protestant and Catholic Irish and Highland Scots in the Maritimes, there are subtle differences in tradition among Highland Scots Catholics in both Newfoundland and Cape Breton, and among the Catholic Irish people of the Avalon Peninsula. Dialectal differences among the rural English of the Newfoundland outports demonstrate the same point.

All of the Atlantic Provinces include Canadian Indians in their rural populations, Malecite in New Brunswick, Micmac in Nova Scotia and the two island provinces, Naskaupi and Montagnais in Labrador. They occupy reserves in the mainland provinces, and are attached to traditional patterns of primary exploitation and seasonal movement elsewhere. Newfoundland has an Eskimo population also in Northern Labrador which has undergone resettlement and centralization in recent years, with consequent changes in its way of life. Numerically the Indians and Eskimos are relatively insignificant, but at the micro-level they constitute locally important elements in the ethnic mosaic. Discrepancies in the percentages of those claiming Indian/Eskimo ethnic origin and those using an aboriginal language indicate, as in the case of Gaelic, that cultural loss has occurred.

The ethnic mosaic is, by definition, a fragmented pattern. In the Atlantic Provinces as elsewhere, it shows signs of erosion. Nevertheless its survival in strength in the cultural geography of the region is a demonstration of cultural resistance. It accounts for much of the richness of the human resource of this part of Canada.

KINSHIP AND SETTLEMENT

Beneath the ethnic macro-structure of the population of the Atlantic Provinces lies a micro-structure of agnatic kinship, indicated by surname and strongly related to pioneer settlement, land tenure, access to the marine resource, and resettlement. It is a structural aspect of settlement both on the land and in the outport, and is characteristic of those areas of *Acadien* French, Southern Irish, and Scottish Highland settlement found throughout the Atlantic Provinces. It is also characteristic of those areas of Newfoundland settled from southwestern England in the seventeenth and eighteenth centuries. Access to the local primary resource base, both land and sea, is firmly rooted in this structure, although it should be recognized that the corresponding affinal kinship structure of society is also always present, exerting a cohesive effect within the individual community and with neighbouring communities. Work on the settlement of Catholic Highland Scots in the Codroy Valley and St George's Bay of Newfoundland between 1840 and 1880 (Delaney and Macpherson) indicates that the flux of secondary and tertiary migration around the shores of the Gulf of St Lawrence, which seems to have characterized at least part of this ethnic group, involved ventures in settlement by brother-in-law more often than brothers. Once a settlement was established, however, and the generations began to multiply, the agnatic principle reasserted

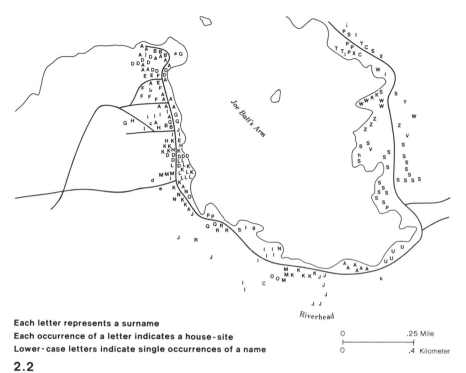

Each letter represents a surname
Each occurrence of a letter indicates a house-site
Lower-case letters indicate single occurrences of a name

0   .25 Mile

0   .4 Kilometer

2.2

Agnatic Kinship Structure, Joe Batt's Arm, Fogo Island, Newfoundland 1968

(from Ralph Brown)

itself to create patterns similar to those shown in Figure 2.2. On the other hand, work carried out by Mannion has demonstrated that the Cape Shore between Placentia and Cape St Mary was colonized in the early nineteenth century by unrelated nuclear families from Ireland, which in the fourth and fifth generations produced agglomerated settlements in each cove, almost exclusively of one surname (Mannion).

Figure 2.2 shows the agnatic kinship structure of Joe Batt's Arm, a settlement on Fogo Island colonized by Newfoundlanders of English and Irish extraction towards the end of the eighteenth century. Initial settlement occurred along the western shore and in the northeast corner of the Arm, where deeper water is found inshore and closer to the fishing grounds just outside the bay, and it is here that the patrilineal descendants of the earliest settlers are to be found: Brown (A), Freake (D), Brett

(F), Higgins (T), Donahue (P), Emberly (C), and Coffin (K). River-head, at the head of the Arm, was not settled till early in the twentieth century. The pattern throughout is one of agnatic kin cells, each representing a discrete piece of original property, once family garden or small holding, but later converted to residential yards to accommodate younger generations as they married. In many instances these kin cells are recognized as neighbourhoods with distinctive names, e.g. Decker's Hill (I), Brown's Point (A), etc. They represent the cores of the units of land tenure and resource use, and constitute one of the most enduring features in the cultural geography.

RELIGIOUS AFFILIATION AND SETTLEMENT

Religious affiliation is an integral and important part of local culture, whether it denotes ethnic differences or not. Historically, and still largely today, it has continued as the dominant factor in political activity, in the provision and form of educational facilities, and in the organization of social life. In specific instances it has even determined retail trading connections between settlements, affected economic development, and influenced the direction of movement in processes of resettlement and centralization. Its role is epitomized in Newfoundland, where, until 1969 education was provided by five separate denominational systems: Roman Catholic, Anglican, United Church, Salvation Army, and Pentecostal.

The structure at all scales is a fragmented one whether the settlement pattern is one of dispersed farmsteads or agglomerated outports. Figure 2.3 shows the arrangement of residences, churches, and schools in Joe Batt's Arm in 1968. Several features of the pattern can be identified as characteristic of the Newfoundland outport, although variations in different outports may be expressed according to the history of settlement.

Congregational segregation is evident in the case of the Catholics and United Churchmen in Joe Batt's Arm; the Anglicans form the ubiquitous congregation. The Catholics living on the southwest corner of the Arm are the result of in-migration from the all-Catholic community of Little Fogo Island which resettled itself in the Arm early in the twentieth century. The Anglicans at Riverhead on the southern shore of the Arm resettled there at about the same time from Barr'd Islands, a smaller community a mile immediately southwest of Joe Batt's Arm, largely to acquire land. The isolated group of Anglicans on the east side of the Arm are the result of recent purchase of land from original Catholic owners.

The pattern, with reference to the present and past locations of churches

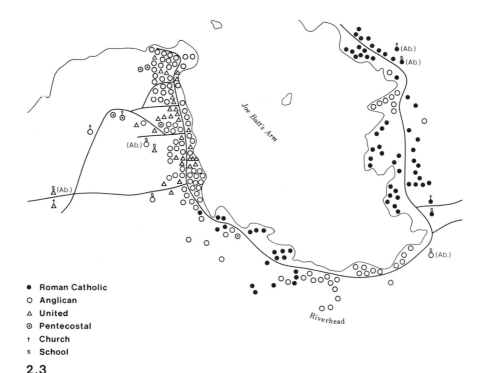

- ● Roman Catholic
- ○ Anglican
- △ United
- ⊙ Pentecostal
- † Church
- s School

2.3

Denominational Structure, Joe Batt's Arm, Fogo Island, Newfoundland 1968

(from Ralph Brown)

and associated schools, and including the Anglican/United Church community of Barr'd Islands is simple *von Theunen*: the Catholic church and school were moved from the northeast to the southeast corner of the Arm to accommodate the Catholics in the southwest; the main Anglican and United Church facilities were located behind the community on the west side, central to Joe Batt's Arm and Barr'd Islands, and a small one-room Anglican school in the southeast corner was abandoned. The Pentecostal church and school had no adherents or attenders from Barr'd Islands, and were located to serve a few recent converts. In other cases, where the physical site and the history of settlement has been different, the churches and schools of two or more denominations are often located close together in a central position and the segregated pattern is lacking. As a consequence, social interaction is likely to be different, and religious affiliation is less likely to be a divisive force in the community.

TENURE AND CONTINUITY: PICTOU

In order to understand the present state of the rural population in the Atlantic Provinces, detailed examination is necessary. This has been recognized in the comprehensive collection of data on property and people in Prince Edward Island and various parts of Nova Scotia and New Brunswick carried out in co-ordinated fashion by federal agencies in collaboration with provincial authorities during the late 1960s. Pictou County, Nova Scotia, was one area so treated, and the following picture emerges from the statistics for an area of some 135 contiguous farms in the western part of the county. It was an area of homogeneous ethnic tradition, having been settled by Highland Scots immigrant settlers and their descendents for almost two hundred years.

In 1968 the farms ranged in size from 9 to 1000 acres; the median farm was exactly 150 acres, including 30–40 improved acres, and the three-quarter percentile farm in the group was just over 200 acres, with 50–60 of them improved. The modal distribution was therefore heavily skewed towards the small farm, as the following figures at 100-acre intervals demonstrate:

0–99	100–199	200–299	300–399	400–499
38	52	25	10	1

500–599	600–699	700–799	800–899	900–1000
1	1	1	1	1

The economy was one of mixed farming, combining arable land, livestock, and woodlot.

Almost every farm had its improved acreage seeded down to hay, and some had a few acres of oats and other field crops. But only 64 farms (47 per cent) retained milk cows with followers; only 15 of these included sheep; only 12 kept pigs. Thus over half the farms were defunct so far as livestock management was concerned. Forty-five farms in the modal 100–199-acre category had an average household income of $3398 per household: only twenty of these farms derived income from the farm operation, averaging $785 per farm; only nine of them had productive woodlots averaging $190 each; only twenty-nine of them had off-farm income averaging $3003 each; thirty-five households obtained an average of $829 per annum from pensions and other transfer payments. Among the twenty productive farms in the sample, only one derived the major part of the family income from the farm: a 71-year-old man living alone on a 170-acre holding which he inherited in 1955. His income was derived from

50 acres of improved land and 23 acres rented in, mostly hay ($900) but twelve acres in special crop ($2800), six milk cows ($500) and followers ($300), and ten pigs ($900). Three men owning farms in the modal category derived substantial income ($4000–8000) from fishing, and in fact headed the most prosperous households in the group.

Of the 135 farms in the block all but two were owner-occupied. The modal distribution of the owner-occupiers by age at 10-year intervals was as follows:

20–29	30–39	40–49	50–59	60–69	70–79	80–89
2	12	22	39	36	14	8

Three-quarters of the farmers were aged 47 and over, half of them aged 57 and over. This distribution, weighted somewhat towards the older age-groups, cannot be interpreted, however, as anything but normal. Tenure of farm property and the general role of succession are sufficient to account for it as virtually universal. So long as there is no threat to continuity of succession in the farm it constitutes no problem. And so long as there is continuity of succession throughout a neighbourhood, the folk tradition – ethnic, religious, linguistic – is safe. It is necessary, therefore, to examine acquisition and family structure farm by farm.

Land in this part of Pictou was acquired by three means: inheritance, purchase from relatives, and purchase from non-relatives. Inheritance, in an area of strong ethnic traditions, can be regarded as the traditionally normal method of succession. Purchase from relatives may also be normal in some traditions, in that it is preferred to gift or inheritance; purchase may also be more or less nominal. Purchase from non-relatives, on the other hand, can be assumed to represent a threat to the tradition of the neighbourhood if it is non-selective and occurring in a free market. In a group of 132 farms for which information exists the pattern was as follows:

Acquisition	No. of Farms	Percentage	Av. age of acquirer	Av. year of acquisition
Inheritance	80	60.6	60	1941
Purchase from relatives	31	23.5	52	1952
Purchase from non-relatives	21	15.9	57	1950

Cross-references for the average age of the acquirer and year of acquisition have been added to the table, and indicate that inheriters are older and have been on their farms longer, while purchasers from non-relatives are younger and are more recent arrivals. The earliest case of inheritance occurred in 1900; the earliest case of purchase from non-relatives in 1935. Clearly, the statistics indicate an on-going process which is acting against inherited succession and the survival of the local tradition, albeit in an early stage and slowly; the average size of the inherited farm was 192 acres, the average size of that purchased from non-relatives 111 acres. Elsewhere in the Atlantic Provinces the process has undoubtedly proceeded to a greater or lesser extent than in this part of Pictou County.

The explanation for the process is to be seen in part at least in family structure, for which data exist for 133 farms. The 'family' on 44 farms (33 per cent of the total) consisted of single persons (unmarried: 13 farms; widowed: 11 farms). If married couples living alone (31 farms) are included, the figure rises to 55 farms (41 per cent); if married couples and unmarried persons with one or two extra adults (17 farms) are added, the figure goes to 72 farms (54 per cent) without a person under the age of twenty-one. In every category the median and average ages of the heads of household are between the ages of sixty and seventy, and virtually every extra adult is a slightly younger sibling or an unmarried son or daughter of middle age. Families with children under twenty-one occupied 61 farms (46 per cent); families with children under fourteen occupied 43 farms (32 per cent). Caution must be used in interpreting this synopsis in terms of process: it could represent a situation that has been repeated over several decades or generations and might be replicated again as family circumstances change and out-migrant sons and daughters succeed to property and return, either as managers or pensioners. What is certain, however, is that western Pictou County had a pattern of family structure in 1968 quite different from that when the traditional economy was in full swing. It is also certain that it does not adequately portray the extreme situations that exist elsewhere in the Atlantic Provinces, particularly in some parts of New Brunswick and Cape Breton Island. Nevertheless, Jackson and Maxwell's study of the Tantramar area of New Brunswick and Raymond and Rayburn's study of land abandonment in Prince Edward Island confirm the case of western Pictou as being representative of a widespread condition.

RESETTLEMENT AND A CHANGING SETTLEMENT HIERARCHY

If the fragmented patterns of ethnicity, kinship, and linguistic or religious affiliation represent fundamental structures in the cultural geography of the Atlantic Provinces, the general condition of the rural population and the cases of western Pictou County and the Tantramar area would seem to indicate that these structures are threatened with fracture and disintegration. Governmental planning and development agencies, both provincial and federal, advocate rationalization of settlement patterns as a solution to what is conceived of as an obsolete population distribution. For the dispersed rural population of the Maritime Provinces the intention is to offer alternatives to non-farming owners in order to release land for consolidation and development. Expansion and rehabilitation of agriculture, woodlot management, and tourism are the goals for rural redevelopment, while resettlement in new urban growth centres and social service centres is the long-term opportunity offered to the rural commuter and pensioner. Housing is seen as the key to the success of such programs of resettlement. Implementation began in Prince Edward Island under the administration of a federal/provincial development corporation in 1970.

Newfoundland has a much longer history of resettlement under governmental direction and encouragement, beginning shortly after confederation with Canada in 1949. Between 1954 and 1965 a provincial centralization policy succeeded in evacuating some 185 communities, involving some 8000 persons. The purpose was to bring people in small isolated coastal communities, many of them on off-shore islands, within reach of adequate social services. In practice the movement was one towards road connections, the period being one of rapid extension and integration of the road system and contraction of coastal shipping services. Individual family movement was therefore minimal in terms of distance and often involved social rather than economic considerations: the presence of kinsfolk or people of the same religious denomination.

In 1965 a joint federal-provincial program of resettlement was inaugurated, originally to promote movement from small inshore fishing communities towards designated growth centres associated with growing deep-sea dragger and seiner fisheries. In practice it meant movement towards fish plants. Later a wider range of designated growth centres and reception centres with social services was introduced. By 1970 a further 18,000 individuals from 126 evacuated communities and 292 non-designated communities had been resettled. It has been suggested that if the policy continues about 17 per cent of Newfoundlanders – 40 per cent of the province's rural population – will eventually move, and a radical redistribution of population and restructuring of the settlement hierarchy should become evident. Already drastic changes have been effected in the major bays of the northeast coast, from Trinity Bay to White Bay, and along the south coast. Again, the movement has been characterized by minimal distance, both linear and social, as is indicated by Figure 2.4a and b.

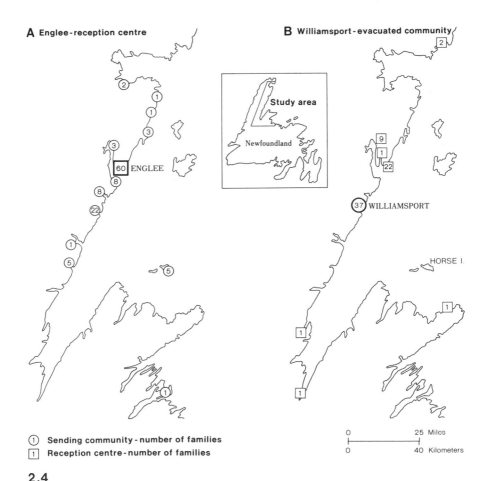

A Englee-reception centre

B Williamsport-evacuated community

Study area

Newfoundland

ENGLEE

WILLIAMSPORT

HORSE I.

① Sending community - number of families
☐ Reception centre - number of families

0   25 Miles
0   40 Kilometers

2.4

Patterns of Resettlement and Centralization: Reception and Evacuation in White Bay, Newfoundland 1965-1970

Figure 2.4a represents the reception of relocating families at Englee, an incorporated community on the east side of the Northern Peninsula which had a population of 941 in 1966. By early 1970 some 300 persons in sixty families had been added, to raise the reception centre out of the rural category for 1971 (population: 1045). All but three of these families left places which had populations smaller than 200 individuals in 1966 and which had been losing population earlier; one third of them came from Williamsport, an isolated community of 151 individuals in 1961, which by the end of 1969 had been completely evacuated.

Figure 2.4b provides a simple demonstration of the process of fragmentation which occurs when communities are evacuated. Of the thirty-seven families involved in the evacuation of Williamsport nine went to the logging centre of Roddickton, while it may be surmised that the rest made minimal moves towards the predominantly United Church settlement of Englee and the largely Anglican cluster of outports at the bottom of White Bay. Actual choices were probably dictated by availability of land on which to reside and access to information networks concerning local and occasional work: both to some considerable extent depend upon kinship and religious affiliation. All but three families moved to places which were in the very small urban category or had joined it by 1971. No long-distance moves to other parts of the province were undertaken to indicate any radical shift in aspirations. The Horse Islands community, on the other hand, with a peak population of 215 in 1956 and totally evacuated by the end of 1969, sent one family to Glovertown, a service centre in Bonavista Bay, three to Marystown, a shipbuilding centre in the Burin Peninsula, and one to the fishery growth centre at Port Saunders on the Gulf coast of the Northern Peninsula. Moves such as these, replicated around the Island, undoubtedly have a cumulative effect on the growth of such places.

Resettlement in the Atlantic Provinces, therefore, would seem to imply a drastic reduction in the numbers of people living in the rural sector of the settlement hierarchy, a necessary rationalization for a demographic situation characterized by depletion in the productive and reproductive age-groups and by the threat of a reduced birthrate. Paradoxically, and despite resettlement in Newfoundland or the Maritimes, the actual number of people in the rural sector of the hierarchy continues to rise. Should this trend reverse, and assuming that the same forces for change are acting generally upon the 'very small urban' sector, further changes in the structure of the settlement hierarchy can be expected. Each change, at the family and community level, represents in turn a loss of immediate access to the primary resource base, terrestrial and marine, and an abandonment of hitherto favoured niches within the physical environment. A further erosion of the cultural heritage seems inevitable.

ACKNOWLEDGMENTS

I wish to acknowledge the aid of a grant from the National Advisory Committee on Geographical Research, and to thank Mr C.W. Raymond for permission to use the socio-economic data on Pictou County.

3 Resource Utilization: Change and Adaptation

W.A. BLACK and J.W. MAXWELL

Fragmentation of the environmental resources has been a critical factor in the use of natural resources in the Atlantic Provinces from the time of the first settlement to the present day. In farming, mining, forestry, and in the fisheries, small operations developed that were highly individualistic; a tradition and pattern of ownership emerged that provided a resistance to change. This legacy from the past has hindered the formation of larger economic units compatible with the modern competitive demand for primary resources. Associated with the concept of fragmentation is another dimension, the concept of scale, which generally emphasize size within a particular sector, for example, comparing large farm units with small farms, the offshore fishery with the inshore fishery, large woods operations with small operations, and so on. The socio-economic characteristics of the sectoral economies have been discussed and operations of scale have been contrasted either with those within the regional framework or with those at the national level.

In the detailed examination of the fisheries and agricultural sectors it has been shown that the impact of fragmentation reached deeply into the socio-economic fabric of both the fishing and farming economies; the discussion of the impact of fragmentation has not been exhausted in this study but rather its forms have been identified. The impact of fragmented resources on the communities of the Atlantic Provinces has contributed to a body of attitudes which, affecting resource utilization, must be considered in the formulation of wise and judicious resource policies of the region. The significance of scale, in a sense, is a more recent phenomenon in the socio-economic planning of resource utilization. The utilization of extensive forest tracts and of large mineral and power resources has been less inhibited by local attitudes because of the need for very large capital outlays and highly specialized managerial and technological skills. These were not readily available within the region, but were drawn largely from the international pool, thus providing a base for operations of vast scale and a freedom from the restrictive characteristics of tradition. This orientation provided a modern thrust in the use of primary resources in keeping with the social and economic realities of the region.

Each section traces the impact of resource utilization on the local economies, the role of change and adaptation, the limitations imposed on policy and regulatory control, and the emerging patterns of resource utilization indicative of a more rational use of the region's resources. The impact of fragmentation and scale on the economic and social development of the resources of the Atlantic Provinces is discussed under the following headings: Fishery Resources of the Atlantic Region; Forest Utilization of the Atlantic Provinces; Agriculture in the Atlantic Provinces; Mineral Resources of the Atlantic Provinces; Energy Resources of the Atlantic Provinces.

Fishery Resources[1]

The historical conception of the fisheries which has continued to the present time is that the Newfoundland, maritime, and foreign vessels concentrated on the cod fishery to the neglect of almost all other species; the schooner fleets were developed to meet the requirements of the salt-fish industry, and the inshore fisheremen of the Atlantic provinces were similarly oriented. The dominance of cod in the economy of Newfoundland is aptly expressed by the phrase 'cod is currency'; to a lesser extent, cod held a similar position in the economic life of the communities that bordered the shores of the Atlantic region. The cod fisheries were prosecuted on a massive scale; the marketed product was salted and the major consuming markets were the Caribbean–Latin American and the European–Mediterranean nations. The universality of cod in the inshore and offshore waters brought about the development of a number of famous fisheries which exploited the common resource and, in recent years, led to the increasing pursuit of new species for commercial development. This study is concerned mainly with the concepts of fragmentation and scale as these apply to the catching-related aspects of the industry.

The progress from the historical conception of a monolithic industry to the multi-species-oriented fresh and frozen fish industry has produced a profound change. The wider range of economic opportunity in the contemporary commercial fisheries has produced greater fragmentation in the catching process: from a diversity of boats and gear to a diversity of activities among fishermen and to a broader and more critical market for the products of the sea. The fishing communities in the past had carved out fishing zones adjacent to the settlements for the exclusive use of their

1   The data for the tables in this section are from DBS Fishery Statistics 1968. Cat. Nos. 24-202, 203, 204, and 205, 1970.

inshore fishermen. The industry depended upon an abundant labour supply; it was an intensive labour-oriented industry. Except for communications by sea, most shore communities, until recent times, were isolated from one another.

The traditional concept of the fishermen's life is that it offered hard work, long working hours, and low pay, but that it was a highly individualistic and independent way of life. The inshore fishery is traditionally a small-scale family operation requiring small capitalization; moreover, the local merchant assured uniformity of product and the great mercantile houses assured the unchallenged supremacy of salt codfish production in the Atlantic region. Nevertheless, it could only provide a subsistence income during periods of good inshore runs of cod. As economic activity in other industries slowed down, the men returned to their home ports and to the sea – a change made easier because of the low capitalization of investment. Over a period of almost 50 years a clear pattern emerges: the number of inshore fishermen declined during years of increased economic activity and rose during periods of depressed economic growth.

Economic opportunities today, as in the past, affect the number of fishermen engaged in the fishery; their numbers fluctuate widely from year to year, from area to area, and from community to community. The larger number of commercial species taken from the sea has provided greater stability in the industry than in the past. The inshore fisherman's operations have become more production-oriented and have moved steadily towards greater capitalization, with larger boats and a diversity of gear and fishing activities. It has become difficult in the deep-sea fisheries to attract young men to spend long periods of time at sea; for the inshore fisherman, to go to sea in the morning and to return in the evening is an attractive feature; furthermore, prosperity in other industries provides out-of-season employment for the inshore man. The inshore fishery is still labour-oriented; in contrast, the offshore fishery is highly mechanized, requiring heavy capital outlays; in effect, it is a big-business operation.

The historic fisheries were each distinctive, yet their structure was remarkably homogeneous in terms of fishing techniques, organization, equipment, and capitalization, and in the exploitation of a common resource; it was remarkably homogeneous in terms of its export product, marketing areas, and marketing organization. The fisheries were structured on a broad regional basis: in addition to the inshore fisheries of the Atlantic region, there were the Newfoundland stationers and the Labrador floater codfisheries – fishing operations conducted off the coast of Labrador, and the Grand Bank fisheries operated mainly by Newfoundland and Nova Scotia fishermen. In recent times there have developed such specialized forms as the lobster, scallop, oyster, and herring fisheries, to name a few. World War II brought a marked reduction in the number of vessels fishing off the Labrador coast and on the Grand Banks. After the war these fisheries were beset by rising costs and inflationary trends, improved economic opportunities in other industries generated by the war, changes in consumer tastes, low market prices for fish in the traditional consuming areas, difficulties in currency convertibility, and an advancing new technology in gear and ships in the state-subsidized foreign fishing fleets. These factors brought about a rapid decline in the stationer fishery, a near-extinction of the Labrador floater fishery, and the end of the world-famous Grand Banks' fleet. Within the industry, the dominating role of salt cod-fish was being challenged by the new fresh and frozen sea products which were meeting the more exacting requirements of the modern European and North American urban markets. To achieve greater efficiency, the fishermen were aided by various kinds of provincial and federal government support programs aimed at modernizing the fishermen's operations. Processors were provided with similar assistance in the construction of new processing plants, in the acquisition of modern fishing gear, and in the construction of new vessels. Provincial and federal governments encouraged the development of strategically located, fish-processing plants in growth centres, integrated into the network of roads and highways communication. The new technology of the fishing industry is oriented basically to prosecuting the fishery in offshore waters. From these developments the scale and volume of the fishermen's operations have increased in efficiency and productivity.

The cod fisheries, which were the basis of economic development and economic well-being and, indeed, responsible for the early settlement patterns in the Atlantic region from Cape Cod to Labrador, were slow to feel the technological revolution that was sweeping the primary industries in the twentieth century. New innovations, which had transformed the northern European fisheries, and by the 1930s the New England fishing industry, were not commonly applied in the Maritimes until the 1940s and 1950s. More specifically, it was the introduction of the deep-sea otter trawlers in the offshore fisheries at the end of World War II that gradually displaced the labour-oriented, salt-cod, dory-schooner operations in order to meet the requirements of the new fresh-fish processing plants; it was the first major change leading in the direction of a capital-intensive fishing operation. National and international pressures on the fisheries, which had brought about the application of scientific methods and technological innovations, set the pace in the catching and processing phases of the industry; these can be expected to set the pace for future technological advances, for the exploitation of underdeveloped species and for the search for new commercial species.

Consumer tastes have always been difficult to assess because sea products have tended to be similar; nevertheless, the number of species of shellfish and finfish for domestic consumption has increased to forty or more. This expansion, aided by modern refrigeration, cold storage, packaging, product standardization, and retail marketing outlets has helped to maintain the freshness of sea products demanded by the urban consumer. Because the contemporary fisheries have been the object of intensive utilization by a large number of primary and secondary producers, the industry has tended to be production-oriented; it has concentrated on production rather than on marketing, so that marketing activities have lagged behind production and marketing has tended to be fragmented. The awareness of the need for change has not abated. The industry, having become technically more stabilized in its catching and processing sectors, may seek greater stability and scale in its supply and marketing phases.

The range of commercial species in the Atlantic region is shown in Figures 3.1 and 3.3. Each provincial fishery in the Atlantic region is structured on a different mix of species, so that the main components have relatively different standing in each province. In 1968 the Atlantic Provinces caught 2,327.2 million pounds (Fig. 3.1). Of this volume, Newfoundland produced about 41 per cent, Nova Scotia 34 per cent, New Brunswick 23 per cent, and Prince Edward Island 2 per cent. Groundfish species accounted for 47.3 per cent, pelagic 49.7 per cent, and shellfish 3.0 per cent. Groundfish led in the Newfoundland fisheries with 65.0 per cent and in Prince Edward Island with 49.1 per cent; pelagic species dominated the New Brunswick catch with 81.8 per cent of the catch. Nova Scotia had the most equitable distribution between groundfish and pelagic-estuarial catches; each amounted to about 48 per cent.

The landed value of the catch in the Atlantic Provinces in 1968 amounted to $102.6 million; groundfish accounted for 43.8 per cent, shellfish, 37.0 per cent, and pelagic-estuarial catches, 19.2 per cent. Shellfish accounted for 44.7 per cent of the landed value in Nova Scotia, 84.4 per cent in Prince Edward Island, and 38.3 per cent in New Brunswick. The landed value of groundfish accounted for 75.1 per cent in Newfoundland, 39.3 per cent in Nova Scotia, 17.8 per cent in New Brunswick, and 10.6 per cent in Prince Edward Island. On a regional scale, the Nova Scotia fisheries contributed about 46 per cent by landed value of groundfish and about 61 per cent of the region's shellfish production; its share of the estuarial catch was about 42 per cent.

Shellfish, which accounted for 3.0 per cent of the landings, contributed

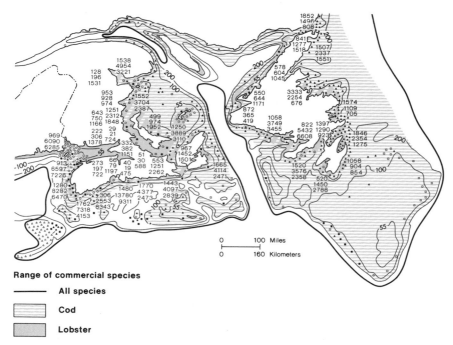

Range of commercial species

———   **All species**

| | **Cod** |

| | **Lobster** |

Location and weight of commercial catch 1963 (includes Quebec catch)

∘ **Herring**

• **Lobster**

∘ **Flounder**

• **Haddock**

∘ **Scallop**

∘ **Cod**

Each symbol represents 1 360 000 kilograms round fresh weight

These species accounted for 80% of the total catch both by weight and value in 1963

1520
3576   **(by fishing districts)**
2353

Total number of fishermen 1968
Total landed value of catch 1968 ($'000)
Per capita value of catch 1968 ($)

3.1

Fisheries

(Sources: A.P.E.C.
        Canada, Dept. of Fisheries
        D.B.S. Fisheries Statistics of Canada 1968)

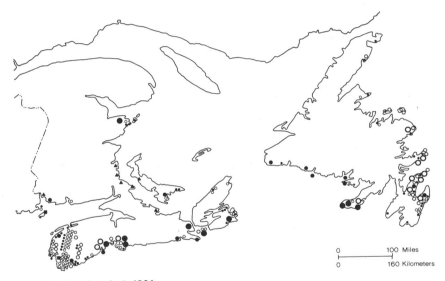

Fresh-fish freezing plants 1964
Raw fish processed (in millions of kilograms per year)

- ·   0.5-4.0
- •   4.5-11.0
- ●   11.5 and over

Salt-fish drying plants 1963
Approximate annual capacity (in thousands of kilograms of dried salted fish)

- ○   Less than 275
- ○   275-679
- ○   680 and over

Other important plants 1965

- ▲   Lobster
- ■   Fish meal
- ▵   Scallops

3.2
Fish Processing Plants

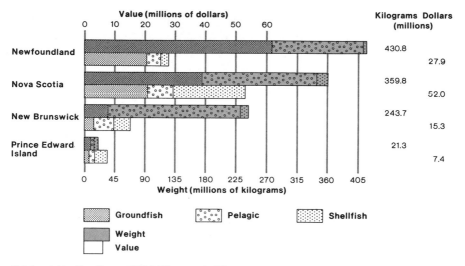

Value (millions of dollars)

Newfoundland   430.8
                        27.9

Nova Scotia   359.8
                        52.0

New Brunswick   243.7
                        15.3

Prince Edward   21.3
Island
                         7.4

Weight (millions of kilograms)

Groundfish   Pelagic   Shellfish

Weight

Value

Total weight of landings : 1055.6 kilograms (millions)
Total landed value : $102.6 (millions)
Livers, viscera, whales, seals and seaweed are not included

3.3

Comparative Landings and Landed Values for the Atlantic Provinces 1968

(Source: D.B.S. Fisheries of Canada 1968)

37.0 per cent of the landed value of fish products of the Atlantic Provinces. In sharp contrast, groundfish made up 47.0 per cent of the landings and 43.8 per cent of the value; and the pelagic catch, 49.3 per cent of the landings, contributed only 19.2 per cent of the landed value. Prior to 1965 groundfish usually made up about three-quarters or more of the landings of the Atlantic region, but the rapid increase in the herring catch has raised the landed value of the pelagic catch from about $12 million in 1965 to $21 million in 1968.

In the historic past, cod was considered the most important fish, both in terms of quantities caught and of landed value; cod holds neither of these positions in the contemporary fisheries of the Atlantic region. The phenomenal growth of the herring fishery from 1964 to 1968 was such that the volume of herring landed in 1968 was almost double that of cod. The leading species are ranked in terms of landed weight: herring (46.8%), cod (23.1%), plaice (8.0%), redfish (6.1%), haddock (3.9%), lobster (1.5%), and scallops (0.6%); these seven accounted for 92.3 per cent of all fish landings. The leading species, ranked in order of landed value, are lobster (22.0%), cod (21.5%), scallops (12.6%), herring (11.1%), haddock (6.6%), plaice (6.0%), and redfish (3.5%); these accounted for 85 per cent of the landed value of fish caught in the Atlantic region. Although the landed weight of lobster and scallops amounted to only 2.1 per cent, the landed value amounted to 34.1 per cent. These commercial species do not rank equally throughout the Atlantic region; thus, the quantities of landings and landed values vary considerably from area to area and according to the mix of species. The number of species exploited throughout the Atlantic region varies considerably: some half-dozen or fewer species are exploited commercially on the northeast coast of Newfoundland and Labrador. On the south coast of Newfoundland about twenty commercial species are exploited. The volume and value of the catch increase markedly southward, reaching a maximum for the Atlantic region on the south and southwest coasts of Nova Scotia where some forty or more species may be taken commercially. It is the southern Atlantic front which provides the resource base for technological innovations and large scale operations; it is in these areas that the highly mechanized phase of the Atlantic fisheries is found.

In the 1952–68 period the rate of increase for groundfish has been persistent – an indication of the present growth of mechanization, marketing and consumption of sea food (Table 3.1). Shellfish landings appear to be in balance with stock replenishments. The greatest increase in the 1952–68 period has taken place in the pelagic component. Wide fluctuations in landings and the seasonal nature of production make it difficult to exert control over supplies; thus it is difficult to define a normal year. An example may illustrate: the 1968 catch of groundfish for Newfoundland was the largest of the past 16 years; the largest catch of herring was taken in 1952 before the phenomenal landings which followed 1966; the maximum catch of shellfish, 27.6 million pounds, was recorded in 1964.

The landings of a given species tend to be unstable, some creating both short-term and long-term instability in supplies – conditions which have a marked bearing on market prices and demand as well as on future marketing prospects.

The inshore fisheries, which accounted for 86.1 per cent of the fishermen, produced 37.3 per cent of the 1968 catch; conversely, the offshore fisheries, accounting for 13.9 per cent of the fishermen, produced 62.7 per cent of the fish that were landed (Table 3.2). Only in Newfoundland is the inshore groundfish production (54.4%) greater than the offshore operations – an indication of the dependence of the inshore fishermen on groundfish stocks; Nova Scotia, New Brunswick, and Prince Edward Island inshore fishermen draw from a more diversified bank of species. A striking feature was the greater production per man in the capital-intensive offshore industry compared with that of the labour-intensive inshore

Table 3.1 Comparative landings by provinces, 1952 and 1968 (millions of pounds)

	Groundfish			Pelagic			Shellfish		
	1952	1968	An. % change	1952	1968	An. % change	1952	1968	An. % change
Newfoundland	443.9	618.0	+2.5	96.5	327.4	+15.0	11.5	4.3	−3.9
Nova Scotia	294.3	380.7	+1.8	68.7	378.0	+28.1	29.4	34.5	+1.1
New Brunswick	47.1	78.8	+4.2	181.0	439.5	+8.9	24.2	19.0	−1.3
Prince Edward Island	11.8	23.1	+6.0	61.7	11.3	−5.1	14.5	12.6	−0.8
Atlantic region	797.1	1,100.6	+2.4	407.9	1,156.2	+11.5	79.6	70.4	−0.7

Table 3.2 A comparison of the inshore and offshore groundfish catches of 1968[1] (fishery landings in millions of pounds)

	Inshore fishery			Offshore fishery			Inshore catch as % of combined catches (%)	Ratio of inshore catch to offshore catch (per man)
	No. of men[2]	Total catch	Catch per man[3]	No. of men[2]	Total catch	Catch per man[3]		
Newfoundland	14,087	336.3	23,874	1,270	281.8	221,890	54.4	1: 9.3
Nova Scotia	5,185	59.2	11,418	1,650	321.5	194,848	15.5	1:17.0
New Brunswick	1,470	7.7	5,238	450	71.1	158,000	9.8	1:30.2
Prince Edward Island	693	7.9	11,400	90	15.2	168,889	34.2	1:14.8
Atlantic region	21,435	411.1	19,178	3,460	689.6	199,306	37.3	1:10.4

1 Source: Economic Services, Department of Fisheries, Ottawa Canada.
2 The number of inshore and offshore fishermen estimated only.
3 Data should be used with caution.

Table 3.3 A comparison of landed value, 1968 (in current dollars)

	Offshore groundfish		Inshore groundfish		Inshore lobster		
	Landed value ($000)	Av. value per man	Landed value ($000)	Av. value per man	No. of men	Landed value ($000)	Av. value per man
Newfoundland	9,299	7,322	11,098	787	6,850	2,423	354
Nova Scotia	17,040	10,326	3,138	605	9,432	10,957	1,162
New Brunswick	2,417	5,372	262	178	3,995	3,992	999
Prince Edward Island	517	5,742	269	388	2,993	5,329	1,781
Atlantic region	29,273	8,460	14,766	689	23,270	22,701	976

Note: The number of fishermen, Tables 3.2 and 3.3, do not necessarily total to give the figures for Table 3.4; fishermen usually engage in a number of fisheries.

Table 3.4 New investment in Atlantic fisheries, 1961 and 1968 (provincial investment in thousands of current dollars)

	1961			1968			
	Amt. ($)	No. of men	Av. per man ($)	Amt. ($)	No. of men	Av. per man ($)	Growth factor
Newfoundland	1,500*	18,756	80	5,937	19,355	307	4.0
Nova Scotia	4,760	12,578	378	17,230	13,108	1,314	3.6
New Brunswick	2,432	6,083	400	3,583	5,942	603	1.5
Prince Edward Island	948	3,464	274	874	3,301	265	0.9
Atlantic region	9,640	40,881	236	27,624	41,706	662	2.9

*Estimated, should be used with caution.

operations: the regional productivity of the offshore fisherman was 10.4 times greater than that of the inshore fisherman. Great variability occurred in the provincial fishing effort which was reflected in both the inshore and offshore per capita catch.

A comparison of the landed values of the inshore fisherman's groundfish catch with the offshore catch shows that the offshore fisherman performed markedly better (Table 3.3). The inshore fisherman's average yields are low – $689 compared with $8460 for the offshore fisherman. When the landed value of the fishery is considered, the positions of Newfoundland and Nova Scotia are reversed: Newfoundland accounted for $7322 per fisherman compared with $10,326 for Nova Scotia. This disparity is due to the fact that the Newfoundland groundfish catch consisted chiefly of cod, herring, redfish, and plaice, all of which yield low returns to the fisherman. Included in the higher landed values of the Nova Scotia catch is the smaller proportion of low-priced species such as cod, plaice, and redfish and a substantially higher proportion of halibut (4.7% @ 48¢ a pound) and haddock (22.1% @ 7½¢ a pound). Thus the average landed value per pound of groundfish in 1968 ranged from a high of 5.3¢ for Nova Scotia, and 3.3¢ for Newfoundland.

Except in Newfoundland, lobster is much more important in the inshore fisherman's economy than groundfish (Table 3.3). Landed values of lobsters ranged from a high of 61¢ to 69¢ a pound in the Atlantic region in 1968. By adding the landed value of scallops, oysters, and clams to the lobster catch, the average landed value for the inshore fishermen would be substantially higher for Nova Scotia, Prince Edward Island, and New Brunswick; whereas, that for Newfoundland would remain much the same.

The increased fishing effort has been marked in recent years by an increasing rate of investment. The actual amount, varying greatly from year to year and from province to province, has grown from about $9.6 million in 1961 to about $27.6 million in 1968 (Table 3.4). In 1968 Nova Scotia, at $17.2 million, accounted for over one-half of the new capital investment of the Atlantic region. On a per capita basis, capital investment has almost trebled, from $236 to $662; investment in the fishing operations has grown steadily. Although Newfoundland had increased its investment in 1968 by a factor of 3.8, it is still trailing Nova Scotia's 1961 level. Nova Scotia, with a per capita investment of $1314, has the most capital-intensive fishing operations in the Atlantic region. However, if this assessment included only full-time fishermen, the per capita figure, would likely be considerably higher.

The level of per capita investment in the Atlantic region is reflected in the type of vessels employed in the fisheries. In sheer numbers, small craft, such as row boats and flats, predominate; large craft are generally located in areas of high fishery returns. The inshore vessels, represented by the under-10-ton class, have increased slowly, by 0.8 per cent over the 1961–8 period; whereas, a phenomenal increase has occurred in both the 10–24.9-ton and the 25-tons-and-over classes amounting to 8.6 and 7.9 per cent respectively. The greatest growth in all classes has taken place in Newfoundland, demonstrating the greater dependence of Newfoundland fishermen on groundfish stocks, both in the inshore and offshore operations. There is marked preference for the versatile 20–24.9-ton class of vessel which has increased in numbers by 86 per cent in Newfoundland and 39 per cent in Prince Edward Island. The proportion of large offshore vessels varies considerably from province to province; for example, about two-thirds of the 150-ton-and-over class are located in Nova Scotia – a figure that has grown from 40 in 1961 to 159 in 1968.

The numbers of new, modern, fishing craft, together with modern catching equipment, are indicative of the substantial amounts of capital that are being spent annually to develop a versatile fleet that is capable of prosecuting the large number of commercial species of the inshore and offshore marine environments; moreover, it stresses the gravity with which the domestic industry views the pressure of foreign fishing fleets in the offshore waters. Because of low returns it has been difficult to attract capital from private sources. Only the large fishing corporations – the modern counterpart of the old fish merchants – with assistance from provincial and federal governments, can provide the necessary capital. The offshore fishery, to maintain a competitive position, must secure large amounts of capital. The number of processing plants in operation has fluctuated narrowly from year to year; this fluctuation indicates that plant closures or marginal plant operations are likely to be related to low market prices for fish products and that many plants have lacked adequate working capital. Two important factors have tended to accelerate the drive towards a capital-intensive fishery: namely, it is recognized that the inshore and offshore groundfish stocks belong to the same common offshore stocks and that these stocks are intensively sought after by fleets of a dozen fishing nations. Any major expansion must come through the expansion of the offshore fishery and by the use of mechanized, deep-sea vessels; further, a greater number of the capable inshore fishermen must be drawn into the offshore operations.

The extent of fragmentation in the Atlantic fisheries is most clearly marked when the distribution of various species and their changing commercial importance is examined; again, it is reflected in the distribution and range of the fisherman's earnings. The distribution of leading commercial groundfish, pelagic, and shellfish species provides a highly fragmented pattern, ranging in extremes from basically poor areas to rich areas, and reflecting an astonishingly wide range in the degree of endowment throughout the Atlantic region.

Cod is the major commercial species of the Atlantic Provinces; it tends to be a minor species in the Gulf of St Lawrence. Halibut and haddock, important minor species on the south coast of Newfoundland, attain major importance in southwest Nova Scotia. Haddock is the most important groundfish species in the Halifax, Lunenburg, and the Maces Bay fishery operations. Plaice rivals cod in commercial importance from St Ann's Bay to Chedabucto Bay of Cape Breton Island; plaice, witch, and yellowtail are important locally on Nova Scotia's south coast. Pollock varies considerably in local importance. Turbot is important as a commercial species only in the fisheries of Newfoundland's east coast. Cusk and hake are of minor importance throughout the Atlantic region. Plaice, redfish, and cod are the three leading commercial species of Newfoundland's south coast; redfish, cod, plaice, and herring, are the leading species in New Brunswick's Gulf coast. Redfish and cod provide the basis for the commercial ground fishery of Prince Edward Island.

The pelagic and estuarial species show a similar fragmentation in coastal distribution. Herring is the most important pelagic species caught; it is of major importance in the fisheries of Newfoundland's south coast, in the Bay of Fundy, and in the Gulf of St Lawrence. Swordfish is the leading commercial species in local areas of Nova Scotia. Tuna is important in the Passamaquoddy Bay area. Salmon is second in commercial importance on the Labrador coast and in Newfoundland's east- and south-coast fisheries.

The mollusc and crustacean fisheries are similarly fragmented and unevenly distributed around the coasts of the Atlantic region. The long duration of the ice-cover and the cold-water period severely restricts and limits the fishery in the Gulf of St Lawrence and along the coasts of Newfoundland; fishing regulations to assure renewal of supplies limit the duration of the catching or impose minimum-size limits. The most productive areas are the southwest and west coasts of Nova Scotia.

The most prolific lobster fishing area in the Atlantic region extends along the southwest and west coasts of Nova Scotia from Shelburne to Digby; this coast accounted for over half (54%) of the landed value of the catch for the province in 1968; the Northumberland coast accounted for about 17 per cent. The main lobster fishery of New Brunswick is concentrated on the Gulf coast; this area yielded almost three-quarters (74%) of the landed values, with the largest catches being taken from the waters of Kouchibouguac Bay; the Grand Manan area, at the entrance to the Bay of Fundy, accounted for 16 per cent of the landed value in 1968. Prince Edward Island's east-coast lobster fishery accounted for about 45 per cent of the landed value of the Province's lobster catch. Approximately 57 per cent of the landed value of Newfoundland's lobster catch was taken from the inlets of Newfoundland's west coast.

The scallop fishery in Nova Scotia competes with the lobster fishery for first place in value of landings: in 1967 lobsters were leading, but in 1968 scallops were ahead. About 98 per cent of the landed value of the scallop catch is concentrated off the southwest and west coast of Nova Scotia, extending from Lunenburg to Digby; Lunenburg, Yarmouth, and Riverport are the main ports of landings for the offshore Georges Banks scallop fisheries; some 50 vessels landed 9.7 million lbs of scallops in 1969. Some 200 inshore boats, equipped with drags, are engaged in the scallop fishery of Northumberland Strait. Scallops in commercial quantities are taken from the protected waters of Port au Port Bay in the southwest coast of Newfoundland, accounting for 85.5 per cent of the Newfoundland catch.

In value, oysters vie with scallops for second place in the Prince Edward Island fishery; oyster culture is conducted in the protected waters of Malpeque Bay; this area accounted for 65 per cent of the Province's oyster production. Bar clams are taken in commercial quantities in various parts of the Atlantic region, particularly in St Mary's Bay, the Annapolis, Cumberland, and Minas basins. The more important soft-shelled clams are taken in commercial quantities from St Mary's Bay, the Annapolis and Minas basins, and from Passamaquoddy Bay.

The harvesting of Irish moss (*chondrus cripsus*) on the southwest coast of Nova Scotia between Cape Sable and Yarmouth, accounted for 22 per cent of the combined value of lobsters, scallops, and Irish moss in 1968. Irish moss is commercially important in Prince Edward Island; the Island's west coast accounted for 68 per cent of the 1968 crop. In 1969 the Nova Scotia–Prince Edward Island producing areas accounted for 72 per cent of the world's supply of *chondrus*. Commercial quantities of dulse are harvested in the vicinity of Grand Manan Island.

The concept of the fragmentation of the fisheries becomes evident when the importance of a commercial species is examined over a period of time. The sixteen-year period, 1949–68, was marked by the phenomenal growth of technological innovation and mechanization in both the catching and processing aspects of the industry, in co-operation among governments, processors, and fishermen, and in the pursuit of new commercial species. The stability, the rise and fall, of a species during this period varies greatly throughout the Atlantic region, emphasizing the capricious nature of risk and opportunity which characterizes the industry.

Redfish, plaice, and halibut rose markedly into importance in the Newfoundland fisheries in 1952; catfish and turbot in 1965. Herring, which had been declining since 1949, moved up sharply year-by-year after 1965. A marked increase in the production of haddock began in 1954; catches have declined markedly since 1964.

In the New Brunswick fisheries the landings of redfish rose markedly in 1965; witch, unimportant prior to 1956, moved upwards in 1959; pollock rose steadily from 1949 to 1960 but declined sharply after 1966. The landings of flounder increased markedly in 1957; its downward trend began in 1963. From its high 1949–56 levels, shad has continued downward; it is becoming an unimportant commercial species. Oysters declined steadily to an extreme low in 1960; since then production has been rising. Scallops, after declining into a trough from 1958 to 1962, began moving upwards in 1963; a sharp increase in landings occurred in 1968. Irish moss, unimportant commercially prior to 1961, moved up sharply after 1966.

In the Prince Edward Island fisheries, flounder, unimportant commercially prior to 1956, rose sharply in 1957 with a sharp decline in 1968. Plaice, a non-commercial species prior to 1952, underwent two periods of expansion: from 1952 to 1956, and from 1961 to the present (1968). Scallops attained economic importance in the Prince Edward Island fisheries in 1964. Although Irish moss was important commercially prior to 1964, its major growth began in 1965. Haddock rose to marked importance in the Island fisheries in 1952; a strong decline began in 1963, reaching its lowest level in 1966.

In Nova Scotia a marked increase in production began with redfish in 1953, witch in 1956, cusk in 1958, and yellowtail in 1961, the latter declining severely in 1968. Prior to 1965 the landings of herring had steadily increased; from 1965 to 1968 herring landings reached phenomenal new records. Flounder production rose markedly in 1957 but since 1963 has returned to pre-1959 levels. Hake has declined greatly from its 1949–54 level and shad has suffered severely in commercial importance since 1964. Plaice underwent two periods of marked growth from 1951 to 1956 and from 1963 to the present (1968). The landings of pollock reached higher levels in 1952 and again in 1958; from 1966 to the present (1968) pollock has declined to pre-1958 levels. Two marked increases in the landings of scallops began in 1953 and again in 1961, but they have been erratic and downward since 1965. The production of Irish moss attained a new level of growth in 1952, and again 1961. The production of haddock in 1968 was about double that of 1949.

An examination of the commercial growth of several species in recent years emphasizes the policy and regulatory control problems associated with the intensive development of new commercial species. The most phenomenal growth in the 1949–68 period has been the herring fishery, which ranked second to cod in quantities landed and fourth in landed value. Massive catches have been taken from the Georges and Banquereau banks; the former was non-existent in 1960, the latter non-existent in 1968. Landings from the Gulf coast of Newfoundland have grown from 6000 tons in 1961 to 145,000 tons in 1968. Production has increased five-fold in both the Gulf of St Lawrence and the Bay of Fundy where herring were already important fisheries. In 1970, a freeze was imposed on the catching capacity of the herring fleet in the Bay of Fundy.

In 1961 scallop beds that could support small-boat operations were discovered in the Northumberland Strait. In 1964, 30 boats were engaged in the fishery; by 1968, however, the number had grown to almost 200 boats landing about two million pounds of scallops. In 1965 queen or snow crab concentrations were discovered in the Gulf of St Lawrence; commercial production began in 1967 with a catch of about one million pounds. By 1969 production had risen to 18.3 million pounds and the number of vessels equipped for the fishery had risen from 15 in 1967 to 95 in 1969. Since 1965 the shrimp fishery has grown, from 6 boats during the 1966–7 season to 40 vessels in the spring of 1969 when two million pounds were landed. Without a history of exploitation of a species in a region, it is difficult to devise adequate regulatory controls to safeguard its exploitations.

A brief comparison of 1968 and 1969 fishery landings illustrates an aspect of fragmentation that arises because of the nature of the resource; in 1969 Newfoundland and New Brunswick recorded increases and Nova Scotia and Prince Edward Island recorded declines. Shellfish landings were about one per cent higher in 1969 than in 1968. In the groundfish landings, cod decreased in relative importance in this period from 48 to 45 per cent and haddock from 8 to 7 per cent. For the deep-sea fleet, from 1968 to 1969, haddock increased from 8 to 10 per cent while cod decreased in relative importance from 28 to 24 per cent. Herring landings decreased by 7 per cent in 1969. In the Atlantic region lobster landings increased by about 7 per cent, whereas scallops decreased by about 15

Table 3.5 Fishermen and landings, the Atlantic region, 1968

	N.S.	N.B.	P.E.I.	Nfld.	Total
Value of landings ($000)	52,069	15,255	7,399	27,855	102,578
Number of fishermen	13,108	5,766	3,301	19,355	41,530
Landings per capita ($)	3,972	2,646	2,241	1,439	2,470
Ratio to N.S. per capita	1.0	0.66	0.56	0.36	0.62

per cent in 1969. Variability in annual trends indicates that, in a short-term sense, supplies of a particular species tend to be unstable; moreover, production is seasonal. These characteristics of the industry have a marked bearing on market demand and prices for seafood, on industry planning, and on governmental policies for catching, processing, and regulatory controls.

Fragmentation of the sea resources, induced by the varying qualities of the marine environment, ranging from a full year's potential operation along the southern Atlantic front, to an open summer season in the Gulf, has a decided impact on the fishing effort as measured by landed values, on the number of fishermen employed, and on fishermen's earnings. Areas of highest per capita landed values have the largest number of vessels either in the deep-sea class or in the 10–24.9-ton class, the largest numbers of inshore vessels and the availability of processing plants to take the fishermen's production; areas of low return tend to reflect a paucity of sea resources and poorly equipped fishermen. Nova Scotia has the highest landed values per fisherman ($3972): in 1968 it was 2.8 times the Newfoundland figure of $1439 (Table 3.5).

Much more pronounced is the fragmentation of fishermen's per capita value of landings among the coastal communities of the Atlantic region. In Newfoundland landed values ranged from $419 per fisherman for the St George's Bay area to a maximum of $6608 for those of Fortune Bay. Newfoundland's ice-free south coast from Cape Race to Cape Ray has been the most productive, with the average landings of about 3950 fishermen with 2.4 times the provincial average, or $3600. About 75 per cent of all offshore fishermen are located in this area.

The per capita value of fish landings in Nova Scotia followed a similar pattern: landed values ranged in average, from $475 to $11,373 per fisherman. The most productive fishing communities extended from Mahone Bay on the east to Digby in the Annapolis Basin; with landed values ranging from $3506 to $11,575; for some 6270 fishermen the average value of landings was $6997. To the east of Digby, along the Fundy coast, fishery returns dropped sharply.

The entrance to the Bay of Fundy is the most productive area of the New Brunswick fisheries; average fisherman's landed values ranged from $3921 to $14,505; for the 968 fishermen, the per capita average was $6291 or 2.3 times the provincial average of $2699.

Prince Edward Island's most productive fishing operations are located on the east and west coasts where per capita landed values averaged from $2579 to $3673.

The unstable history of sustained operations on a large number of potential commercial species and the low-level of income to be derived from much of the fisheries suggest the difficulties inherent in determining potential supply and indicate that in a labour-intensive industry too many fishermen may be seeking too few fish. The commercial pursuit of a new species is likely to follow a pattern already familiar to the region – a rapid rise in landings, followed by a plateau of variable duration; next, a rapid decline that may taper off to economic unimportance. The supplies of lobsters, scallops, and oysters are inelastic – a situation that is likely to be true for clams, crabs, and shrimps. For pelagic and groundfish species, catches seemed to be approaching levels where added pressure would aggravate the supply shortage – North American consumption of fish products has been growing from 5 to 6.5 per cent a year. Reduction of the number of inshore fishermen is hampered by the widespread inequalities of education and income and the lack of opportunities in other industries in the Atlantic region; moreover, without much greater international cooperation from the fishing nations to conserve the offshore fisheries, stabilization of commercial operations becomes difficult.

The concepts of scale and fragmentation as illustrated in the structure of the fisheries are likely to be greatly modified or intensified by several serious factors affecting the fishery resources of the Atlantic region. Basically these are: pollution, depletion, and stabilization.

Because shellfish may be the carriers of infectious bacteria, the contamination of the water by domestic sewage has resulted in the closure of shellfish beds in the Maritimes. The number of areas closed to fishermen has grown from 48 in 1940 to 183 in 1970. Of these closures, 50 were in Prince Edward Island, 42 in New Brunswick, and 91 in Nova Scotia; about 25 per cent of the oyster beds and about 20 per cent of the clam beds are closed, closures continue, and few beds have been re-opened. The amount of untreated sewage continues to rise: about 43 per cent of the regional population of the Maritime Provinces is served by public sewers without any treatment facilities; only 5.5 per cent of the population are connected to public sewers with treatment facilities, the remaining 51.5 per cent being served by septic tanks, storm sewers, drainage ditches, and 'back-houses.' Contaminated water intensifies the fragmentation of a high-cash-return resource, seriously jeopardizing the economic activity of inshore fishermen.

Polluted waters contain a serious economic threat to the future operations of fish processing plants. Most fishing communities occupy protected harbours within bays and estuaries which provide the best locations for fish processing plants. Domestic and industrial wastes from village and town and chemical wastes from forest and farm lands are carried by streams and rivers to discharge into coastal bays and estuaries. Both industry and government demand that processing plants use water having the same quality as drinking water to assure strict sanitary control in the processing of sea food. The amount of water required varies with the size of plant: up to 200,000 gallons a day for a small plant and over one million gallons for the largest plants. A shortage of clean fresh water has caused many plant operators to turn to sea water. This need raises a dilemma: as urban outfalls are extended seaward, polluted water is extended seaward; moreover, suspended material is transported in all directions into every corner of an estuary by surface currents and prevailing winds. Thus the supply of clean seawater at reasonable piping distance and cost for plant operation can be seriously imperilled. This potential threat hangs over an industry whose annual production from processing plants of the Atlantic region is valued at some $125 million, almost all of which is exported to the United States and Europe.

Intensive fishing pressures from all the fishing nations of the Northwest Atlantic have resulted in foreign landings in 1969 being 14 times the 1958 catch. The Canadian share of the groundfish catch from the Atlantic east-coast banks has fallen from about one-half in 1960 to about one-third at the present time. Although the total aggregate of catch is higher because of the increased fishing effort, productivity of the Canadian inshore fisherman per man-day has been cut in half. The declining catch was evident in pre-World War II figures: the average catch per fisherman of the Labrador floater codfishery in the 1941–50 period was down 32 per cent from the 1931–40 period. The fleets of a dozen fishing nations are considered to be primarily responsible for the declining inshore catches.

The International Commission for the Northwest Atlantic fisheries noted that exploitation of sea resources on a massive scale was the major cause for the decline in major species caught in the Northwest Atlantic in 1970, and that such heavy exploitation in many cases exceeded the point of maximum sustainable yield. Declines occurred in cod, redfish, herring, haddock, and flounder. These five accounted for more than 90 per cent of the total catch of all species taken in the Northwest Atlantic in the past decade. The declining catches are a cause for grave concern: the maximum sustainable yield for cod is thought by some to have been passed in 1968; others believe the catch is within 10 or 20 per cent of the maximum sustainable yield. Redfish appear to be within 15 per cent of the maximum sustainable yield. Haddock landings have declined severely to about one-fifth of the catch taken a decade ago. The decline in landings of red and silver hake and yellowfin tuna has caused international concern. The high seas fishing fleets have increased the catch of the Atlantic salmon to a point that is causing national concern for stocks which originate in streams of the Atlantic region. There is firm evidence that herring catches have reached the maximum sustainable yield in the Bay of Fundy. Because of mercury contamination, fishing for swordfish has been halted. The landings of lobster and scallops are considered close to the maximum sustainable yield consistent with conservation of stocks; moreover, the supply of scallops from the Georges Banks declined in abundance in 1969 to one-fifth of that in 1961. Maritime oysters are slowly recovering from a devastating bacterial attack suffered from 1955 to 1960 when, in 1958, commercial production dropped to about one-quarter of the 1950 landings. It is also believed that the maintenance of the ICNAF quota for seals in the Atlantic region will cause a progressive depletion of seal stocks.

The stabilization of the fisheries is both a national and international problem. Severe depletion of haddock stocks brought about an agreement (1969) that banned the fishing of haddock on the Georges and Brown banks during the spawning season, March and April. The International Commission for the Northwest Atlantic fisheries has recommended a closed season for red and silver hake in certain areas. An international treaty, aimed at scientific management of tuna stocks of the Atlantic Ocean, came into force in 1969. An immediate freeze on the catching capacity of the Canadian Atlantic herring fishing fleet in the Bay of Fundy was imposed in 1970. To maintain salmon stocks, Canada recommended that netting of salmon outside national fishing limits should be prohibited; however, the Danish fishing of salmon in international waters was upheld by a majority of ICNAF members in 1971. Lobsters and oysters are subject to strict regulatory control; the offshore lobster fishery, which has been actively prosecuted by foreign trawlers, was opened to Canadian fishermen on a trial basis for the first time in August 1971.

In order to strengthen the domestic fishing economies, the federal government has moved in several directions. Canada and the United States have recently negotiated an agreement on reciprocal fishing privileges following the establishment by each country of exclusive fishing zones. In 1970 the Salt Fish Development Corporation was established; its immediate objective was to improve and stabilize the earnings of the primary producers of salt codfish; its long-term objectives were to achieve an orderly phasing out of the salt-cured product. An ecological approach has been considered in the new reorganization of the Canadian fisheries through amendments to the Fishery Act, amendments to the Coastal Fisheries Protection Act, amendments to the Territorial Sea and Fishing Zone Act, and the Canada Water Act. The 12-mile exclusive fishing zone was established in 1964. In 1967 straight baselines were drawn from headland to headland, with the 12-mile territorial limit extended beyond the headland–baseline boundaries. These amendments to Canada's territorial sea, which replaced the old three-mile concept were approved by Parliament in 1970. In order to conserve, regulate, and manage the fisheries in the Gulf of St Lawrence and the Bay of Fundy and to bring these water bodies under Canadian jurisdiction, fishery closure lines were drawn across the entrances to the Strait of Belle Isle, the Cabot Strait, and the Bay of Fundy; the closure lines came into effect in 1971. Similarly, Canada's anti-pollution programs were extended to these areas under the anti-pollution provisions of both the Fisheries Act and the Canada Shipping Act. The establishment of exclusive Canadian fishing zones enables the Canadian government to exercise jurisdiction over marine resources so that the resources can be managed and the environment protected for the benefit of the Canadian people. An exceedingly difficult socio-economic problem associated with stabilization is the pressure for the domestic industry to become capital-intensive in periods of prosperity and labour-intensive in periods of retarded economic growth. At the international level a decisive step is required, namely, to obtain co-operation and co-ordination among the fishing nations in order to manage the marine resources efficiently for present and future generations.

Forest Utilization[2]
Environmental features and severe climate fragment extensive forest stands to an extent that, in Newfoundland–Labrador particularly, forests appear from the air as mosaics of irregularly shaped patches of variable sizes interlaced with lakes, swamps, rocks, and barrens. Approximately 50 per cent of Newfoundland and 60 per cent of Labrador consists of non-productive forests; whereas, about 13 per cent of New Brunswick may be classified as unproductive. The potential yield of the productive forests, however, depends mainly on human factors to devise adequate regulatory controls and management practices to grow, manage, and har-

2   The data used for the tables in this section are from 'Forestry in the Atlantic Provinces,' *Background Study*, No. 1, Atlantic Development Board, 1968.

vest forest crops. The need is apparent: paper, the basic export product derived from the forests, is in world competition for export markets.

Productive forest land of the Atlantic Provinces (Figure 3.4), comprises about 47.2 million acres or about 7.7 per cent of the Canadian total; while the 35.6 million acres of non-productive forest land represent 7.4 per cent of the Canadian total. The Atlantic Provinces are estimated to contain 331.4 million cords of merchantable pulpwood, a figure that represents about 10.4 per cent of the Canadian total; merchantable saw timber is estimated to contain about 101.6 million cords or 2.7 per cent of the Canadian total. The distribution of productive forest land among the Atlantic Provinces is shown in Table 3.6.

The hardwood–softwood content varies considerably in the structure of the merchantable forests of the region: in Newfoundland, pulpwood stands may contain about 16 per cent hardwoods; whereas, in the older, mature stands on the mainland, hardwoods may constitute as much as 34 per cent. The growing stock also varies widely throughout the region; in Newfoundland it may exceed 90 per cent softwood and contain less than 10 per cent hardwood; in Nova Scotia it averages about 33 per cent hardwood and 67 per cent softwood, and in New Brunswick about 30 per cent hardwood and 70 per cent softwood. The region is basically a pulpwood timber source for the pulp and paper mills.

The present annual harvest of about 500 million cubic feet or 3.1 million cords (@ 128 cu feet) is about the same as in the past 20 to 50 years – a trend brought about mainly by the sharp decline in lumber production, by the declining use of wood for fuel, and by improved forest management practices. Only in Newfoundland has forest depletion continued to increase. Through the nineteenth and twentieth centuries the mature forest stands were considered valuable saw-timber reserves composed mainly of red spruce and white pine and the more valuable hardwood species; these species mature in 120 to 200 years. The gradual depletion of choice saw-timber stands in accessible areas, brought about through the extension of farming and through the demands of domestic and international trade requirements, had made numerous areas more favourable for the fast-growing and fast-maturing species such as balsam fir, white and black spruce. These short-lived species, maturing in 60 to 80 years, are economically more desirable for the pulp and paper industries than the slow-maturing species. As merchantable timber declined for lumber purposes, that for pulpwood increased, and thus in the early 1950s pulpwood timber production, for the first time, surpassed the cut for saw-timber. The gap between the two has continued: in 1965 saw-log material constituted less than one-third of the pulpwood cut (Table 3.7).

Forested land (includes productive woodland and burnt-over areas)
Percentage by volume for forest districts

	Softwood	Hardwood
	Over 80	20 or less
	75-80	20-25
	70-75	25-30
	65-70	30-35
	60-65	35-40

Bog and barren

Other areas (chiefly agricultural lands)

Estimated volume of timber by forest districts (hundred millions of cubic feet)

———— Forest district boundary

20   Small material (4-9" DBH*: suitable for pulpwood)

10   Large material (10" and over DBH: suitable for sawtimber)

Production of major saw mills 1964 (millions of board feet per year)

∘   1.5-1.9

∘   2.0-2.9

○   3.0-5.9

○   6.0 and over

Pulp and paper mills 1964

Pulp	Paper	Pulp and paper	Production (thousands of tons per year)
•	▲	■	Less than 100
•	▲	■	100-299
•	▲	■	300 and over

* Diameter breast height

3.4

Forestry

(Sources: A.P.E.C.
        Nfld. Dept. of Mines, Agriculture and Resources
        N.S. Dept. of Lands and Forests
        N.B. Dept. of Lands and Mines
        A.D.B. Forestry in the Atlantic Provinces 1968)

Tenure of productive forest land, 1963*
(in thousands of acres)

Area	Provincial Crown				Federal Crown		Privately owned				Total PFL†
	Occupied	Vacant	Total	% of PFL†	Total	% of PFL†	Farm woodlot	Other	Total	% of PFL†	
Atlantic Provinces	24,029	5,747	29,776	62.3	338	0.7	2,881	14,136	17,017	36.1	47,131
Labrador	12,300	1,062	13,362	100.0	—	—	—	—	—	—	13,362
Nfld. (Island)	4,324	2,688	7,192	86.5	—	—	20	1,098	1,118	13.5	8,310
P.E.I.	—	2	2	0.4	3	0.6	267	248	515	99.0	520
N.S.	747	1,425	2,172	22.5	20	0.2	1,363	6,096	7,459	77.3	9,651
N.B.	6,658	390	7,048	46.1	315	2.1	1,231	6,694	7,025	51.8	15,288

Source: DBS, *Canadian Forestry Statistics, 1962.*
*Statistics for Figure 3.4.
†Productive forest land.

Table 3.6   Productive forest acreage of the Atlantic Provinces, 1963
(in millions of acres)

Newfoundland		Nova Scotia	New Brunswick	Prince Edward Island	Total
Island	Labrador				
8.31	13.36	9.65	15.29	0.52	47.13

Table 3.7   Forest depletion, Atlantic Provinces, 1965
(in 000s of cords, at 128 cu feet per cord)

	Sawlogs and bolts	Pulpwood	Firewood	Others	Total
Newfoundland	46.9	643.9	132.8	23.4	847.0
Nova Scotia	332.6	372.4	88.6	8.8	802.3
New Brunswick	428.2	890.3	116.2	48.0	1,482.7
Prince Edward Island	11.7	38.3	23.4	3.9	77.3
Atlantic region	819.4	1,944.9	361.0	84.1	3,209.3

Note: Data for some provinces are estimates only.

Current depletion in 1965 showed that New Brunswick, followed by Nova Scotia, had the greatest production output in two important categories: timber cut for constructional purposes and timber cut for pulpwood purposes (Table 3.7). Newfoundland's forest production has been oriented to meeting pulpwood and firewood requirements. A much greater part of the Newfoundland output entered pulpwood production, a fact that emphasizes the greater dependence on the smaller balsam fir and black spruce species; whereas, a much greater part of Nova Scotia and New Brunswick output entered sawlog production. Most firewood used by the small coastal outports of Newfoundland is taken from the fisherman's three-mile limits bordering the shore. In other areas of the region, much of the firewood comes from farm wood lots, private wood lots, and as a by-product of sawmilling. Timber used for firewood is in competition with other uses; its use can only decline when other fuels can be offered at more competitive prices. In many woodland areas adjacent to villages, rural communities, and farms, and particularly on the 3-mile coastal strip of Newfoundland, forests do not grow to maturity; trees are cut for firewood when they reach a height of 8 or more feet.

The most significant factor, after environmental quality, affecting fragmentation of the resource base and the scale of operations, is land tenure. The tenure of occupied productive forest land in the Atlantic Provinces is shown in Table 3.8. Of the productive forest lands in 1963, Provincial Crown controlled 63 per cent or about 30 million acres, and private holders controlled 36 per cent or about 17 million acres, of which latter almost 3 million acres or 17 per cent are in farmers' wood lots; the remaining 14 million acres or about 83 per cent are held largely by the pulp and paper companies and by large numbers of private holders. Although the pattern of tenure varies considerably throughout the Atlantic region, it has an important bearing in the utilization of timber. Where forest land has been consolidated into large blocks, either through freehold or leasehold, large wood-base industries are assured adequate supplies of low-cost wood over long periods of time. Numerous private holdings prohibit any effective management or regulation of forest land, or even a systematic harvesting of stands and the restocking of potential forest land.

Sawmill operations throughout the Atlantic region depend on logs taken from Crown lands, from pulp and paper company concessions, from farm woodlots, and from owners of small holdings. Most leases are of short-term duration, 2–5 years, while many are on a year-to-year basis; few exceed 5 years. Not many operators own their own stands; thus, the chance for increasing production is slight. Even large mills, exceeding 2 million board feet of annual production, have no more than a 2–5-year supply of timber. Across the Atlantic region the pulp and paper companies hold a large share of the potential saw-log supply, through either leasing or ownership. Thus, most large sawmills depend on the goodwill of the pulp and paper companies and of government departments for their raw material.

The large number of sawmills throughout the region is related to the numerous small private holdings of timber land; these mills engage in custom sawing or lease timber stands for clean cutting or for the purchase of logs. In 1965, Nova Scotia was reported to have about 200 active sawmills exceeding 100,000 board feet production of lumber, of which 31 produced more than 2 million board feet. Of New Brunswick's 279 sawmills, 48 produced more than 2 million board feet. Of the 1100 licensed mills in Newfoundland in 1966, only one produced over 1.0 million board feet; however, one company, controlling three mills, produced over 2 million board feet. Most small mills are push-bench operations, operating seasonally and on a part-time basis; most are two-man operations where production up to 50,000 or more board feet is cut, usually, for local use. Most communities may have from two to six small mills functioning – a practice which fits into a subsistence economic pattern existing in the more isolated areas of the Atlantic Provinces. Land tenure practices, the general provincial government policies to commit Crown land timber to the pulp and

Table 3.8 Tenure of occupied productive forest land, 1963
(in 000s of acres)

	Provincial		Federal Crown	Private			Total private	Regional total
	Occupied	Vacant		Farm	Other			
Newfoundland								
Island	4,324	2,868	—	20	1,098		1,118	8,310
Labrador	12,300	1,067	—	—	—		—	13,367
Nova Scotia	747	1,452	20	1,363	6,096		7,459	9,678
New Brunswick	6,658	390	315	1,231	6,694		7,925	15,288
Prince Edward Island	—	2	3	267	248		515	520
Atlantic region	24,029	5,779	338	2,881	14,136		17,017	47,163

paper companies, and the policies of the pulp and paper companies to retain saw-log material for pulpwood purposes have provided serious obstacles for the develoment of a lumber industry in the Atlantic Provinces.

In Nova Scotia four pulp and paper companies hold freehold leases to 1.6 million acres and, from the Crown, an additional 1.4 million acres are leased, representing about one-third of the productive forest land of the Province. The sawmill companies hold, either through freehold or lease, 434 thousand acres or about one-eighth of the productive forest land. In New Brunswick seven pulp and paper companies own 2.2 million acres of freehold forest land and lease an additional 5.7 million acres from the Provincial government or 52 per cent of the province's productive forest land. The sawmilling industries control, either through freehold or licence, about 750 thousand acres or about 5 per cent of the timber land. On the Island of Newfoundland about 60 per cent of the productive forest land or over 5 million acres is owned or controlled by two pulp and paper companies; about two-thirds of the Crown land or 1.9 million acres is committed to a third company. Of the 13.4 million acres of productive forest land in Labrador, the Crown holds about 4.8 million acres and of the four companies, holding about 8.6 million acres, two operated on the Island. Most timber cut for lumber has come from the fisherman's three-mile limits or is leased from Crown lands.

Fragmentation of licensed lands, the long distance to transport wood to mills, and the fragmentation of productive stands imposed by the environment add to the cost of utilizing the more distant and inaccessible stands. To these may be added costs incurred through fragmentation of cutting operations: cutting may be taking place in several or more stands in the course of the year where yield, accessibility, haulage, volume of cut, distance and the cost of constructing and maintaining logging roads may vary greatly from stand to stand. Consolidated stands, however, provide a base for integrated management practices in order to achieve maximum sustainable yield, but on the large pulp and paper companies' concessions, scientific forest management is practised little.

Private holdings fragment forest stands to an unusual degree: Nova Scotia reported in 1966 some 49,500 separate ownerships of forest lands ranging from 50 to 1000 acres. Although the number of small private holdings has declined, the consolidation of forest lands into larger holdings in Nova Scotia by the pulp and paper companies has continued. Of some 8 million acres of New Brunswick's private productive forest land more than half (or about 56 per cent), that is 4.5 million acres, is privately owned and in small holdings; the remainder is in large freehold properties. Most of the woodland in Prince Edward Island is privately controlled, either by farmers or absentee owners. Numerous private holdings involve a complexity of private arrangements which, together with increasing absentee ownership, make it difficult for wood industries to obtain timber concessions. It might be expected that on farm woodlots or small private holdings management could be more intensive to produce higher yields; however, on most small holdings, forest stands are often of poor quality. On large freehold properties there is a higher level of stocking, less unstocked land, and a higher value of timber per acre than on small holdings. Thus, the fragmented nature of land tenure makes it exceedingly difficult to apply effective controls to attain maximum sustainable yield, and, in a scientific sense, no forest management is practised either in farmers' woodlots or in small private holdings.

The control of forest stands through the existing land tenurial system makes effective management of small private plots difficult and area control of forest lands impossible to apply. Merchantable forests account for 73 per cent, young growth 19 per cent, and unstocked land 8 per cent of the forest areas of the Atlantic Provinces; the distribution of each varies greatly from province to province. For spruce and balsam fir, having an 80-year rotation, merchantable forest should constitute 50 per cent, young growth 45 per cent, and unstocked 5 per cent; only Newfoundland approaches this pattern with merchantable stands accounting for 54 per cent and young growth for 45 per cent; thus there is considerable underutilization of merchantable stands in the Atlantic Provinces with not enough young growth for replacement when the mature forests may be harvested. The allowable cut of both softwood and hardwoods for pulpwood purposes for Nova Scotia, it is estimated, can be increased from 94 to 117 per cent, for New Brunswick from 139 to 370 per cent. Newfoundland, with a third mill would bring the annual cut close to the allowable limit. The present allowable cut in Labrador can be doubled – the linerboard operation at Stephenville is expected to draw 550,000 cords of pulpwood a year from Labrador. The present annual cut in Prince Edward Island can be increased by 50 per cent. Increased production on this scale would depend on the development of export markets; moreover, there is sufficient saw-timber in the Atlantic region to support an industry at least twice the current size – the main problem is one of distribution of suitable saw-timber. Underutilization of both hardwood and softwood stands varies considerably in the region; it is related to demand, to distance, and to accessibility, as well as to the much broader and more incisive factor of land tenure.

Provincial governments have exerted very little authority over the management of either their Crown lands or private forest lands. There is no effective management program to grow timber on an 80-year rotation for the pulp and paper industry or to relegate saw-timber production to a secondary role. Only through integrated utilization can suitable saw-timber from the pulp and paper companies' concessions be directed to sawmills – only then, under expanded production, would marketing constraints become important. The present system of forest utilization has evolved in a social, institutional and political climate that eschewed any form of restraint. To bring about the increases suggested, would require changes in the existing tenure pattern that may be strongly resisted and slow to materialize. The development of scientific forest management practices is likely to be acceptable in the long-term interests of government and industry in order that management of the pulp and paper industries remain solvent in the highly competitive export markets for paper products.

Agriculture[3]

The Atlantic Provinces' agricultural industry entered the 1970s facing major challenges. The long-term decline that characterized the industry for most of the century had accelerated in the decades of the 1950s and 1960s, and the impact of new high-cost technology had created uncertainty in many of the traditional farming enterprises.

In spite of this decline, agriculture has made, and continues to make, an important contribution to the regional economy. In 1961 agricultural exports for the Atlantic region amounted to $411 million or about 38 per cent of the net value of the Atlantic output, that is, agricultural exports were in second place, preceded by pulp and paper in first place (40%), and followed, in third place, by fishery products (19%). Within the provincial structure, the importance of agriculture has varied considerably in the Atlantic region; for example, agriculture traditionally has played a minor role in Newfoundland's economic history compared with fisheries; in Nova Scotia the fisheries and forestry have always held leading roles; and, in New Brunswick, forestry. Of the Atlantic Provinces, Prince Edward Island has been the most oriented towards agriculture: the net value of production from agriculture in the 1941–62 period, amounted to 26.4 per cent for Prince Edward Island, whereas it amounted to 7.6 per cent for New Brunswick and 6.1 per cent for Nova Scotia.

The three Maritime Provinces failed to match the gains made at the national level in farm productivity: in terms of physical volume, agricultural production increased during the postwar period, 1946–65, by only 3 per cent in New Brunswick, 16 per cent in Prince Edward Island, and

3 The data for the tables in this section are from 'Census of Canada, 1966' vol. III, Cat. Nos. 96–601, –602, –603, –604, –605, 1968.

20 per cent in Nova Scotia; in the same period, the national output had expanded by 45 per cent. The gross value of agricultural production in the 1961–5 period was $3459 for the farm worker of the Maritime Provinces compared with $5512 for Canada, or about 63 per cent of the national figure. The net value of agricultural production from 1950 to 1962 at the national level rose about 30 per cent; for the Maritime Provinces the net value declined by about 26 per cent.

The agricultural labour force had been falling at a faster rate than elsewhere in Canada: in 1941 about 26 per cent of the labour force was engaged in agriculture, but by 1961 the figure had fallen to 6.3 per cent compared with a decline in the labour force of Canadian agriculture for the same period from 25.7 per cent to about 10 per cent. In effect, reliance on agriculture had fallen from second place in 1941 to fourth place in 1961.

Low farm income had an adverse effect on farm labour: average net farm income increased in Canada by 59 per cent in the period from the early 1940s to the early 1960s, compared with a decline of 5 per cent in the Maritime Provinces. By the late 1950s, average net farm income in the Maritimes was only $1026, or less than 44 per cent of the average Canadian figure of $2344. Furthermore, total farm-capital investment per worker in the Maritimes was only 48 per cent of the national figure.

The changes taking place in the agricultural sector of the economy are part of a complex phenomenon that has occurred in advanced nations: in the Atlantic region the phenomenon is more pronounced; its effect on the other sectors of the economy is more marked. As farming entails a body of attitudes and practices, it has produced a way of life basically stable and conservative, slow to adapt to changing circumstances, or to adopt new technological innovations. These characteristics are more marked in the farming communities of the Atlantic region, largely, perhaps, because the agricultural areas are fragmented and isolated and lack the cohesiveness and dimensions of the agricultural communities of the St Lawrence lowlands or the West; and perhaps because the region lacks free access to the immense mega-urban markets of the northeastern United States. Agricultural areas are highly fragmented, varying in size from the larger agricultural areas of Prince Edward Island, the St John's River Valley, and the Annapolis Valley, to the small fenced-in garden patches of the Newfoundland outports.

Areas of good farm land are generally limited to river valleys and localized areas of the coastal plain. These fragmented areas are separated by rocky highlands and ridges, rough topography, barrens, thin and porous soils. Agricultural crops are more demanding and their physical require-

ments are more restrictive than those of forestry. Under these physical requirements the climatic regimen is an important parameter, limiting the
range of crops to be grown in an area. Within the limits circumscribed by
environmental factors, effective land use – its productivity and management – is determined by human choices which, in turn, are influenced by
social and economic forces of society. Farm-land patterns are shown on
Figures 3.5 and 3.6.

In the Atlantic region the ratio of the total land area in agriculture to the
total provincial areas has been declining; in 1966, 3.7 per cent of the Atlantic region was in occupied farms – a decrease of about 39 per cent since
1941 (Table 3.9). Undoubtedly the decline is partly related to an increasing abandonment of small-farm operations and to a shift to larger units.

A striking feature of regional agriculture is the low proportion of improved farm land, about 37.0 per cent, to the total farm average; moreover,
improved land has decreased by about 34 per cent from 1941 to 1961,
but from 1961 to 1966 the decrease was 6.3 per cent, emphasizing the
persistent decline in the importance of farming in the region. A comparison of improved and unimproved land in the Atlantic Provinces shows that
Prince Edward Island, at 61.5 per cent, has the highest ratio. In fact Prince
Edward Island's improved farm-land ratio is higher than the Canadian
ratio of 59.9 per cent and the 60.7 per cent of central Canada (Figures
3.5 and 3.6).

In the period from 1961 to 1966 the absolute number of farms dropped
from 33,391 (1961) to 26,393 (1966) – a decline of 20.9 per cent; moreover, each province, in varying degrees, participated in the decline (Table
3.10). Although commercial farms showed a regional gain of 1.1 per cent,
the number of these farms declined in Nova Scotia and New Brunswick;
a small gain was recorded in Newfoundland, while a gain of 13.8 per cent
took place in Prince Edward Island. The Island in 1966 had the largest
number of commercial farms in the region and its rate of increase was
double the Canadian average of 6.9 per cent.

The 1966 Census defined a commercial farm as one having sales of farm
products, during the 12-month period prior to the Census, of $2500 or
more; thus a farm below $2500 in sales may be defined as non-commer-
cial although it may, in fact, contribute substantially to cash income. In
absolute numbers this type of farm accounted for 64 per cent (16,959) of
all farms (26,393) in the region. The non-commercial farms have declined
by about 29.5 per cent from 24,155 in 1961 to 16,959 in 1966 with the
greatest decline of 33.3 per cent having occurred in New Brunswick and
31.8 per cent in Prince Edward Island; presumably the greatest consolidation of farm land was taking place in those provinces (Table 3.10). The

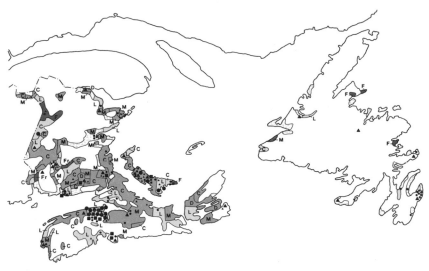

Types of commercial farming 1961
Crop farming

F	Field crops*
Fr	Fruit and vegetables*
M	Mixed crops*

Livestock farming

D	Dairy*
L	Livestock, dairy emphasis**
C	Cattle, hogs and sheep*
M	Mixed livestock

Mixed farming

C	Mixed with crop emphasis***
L	Mixed with livestock emphasis***
	Scattered farming (less than 10 farms in census subdivision)
	Non-farming areas

* 70% or more farmers received most of their agricultural income from sales of specified products

** Over 70% of the farms qualify as livestock type with at least 50% qualifying as dairy type

*** Between 50% and 70% of the farmers received most of their agricultural income from sales of specified products

Food processing plants 1965 (only those engaged in export trade are shown)
Major Minor

■	·	Meat products
▲	·	Dairy products
●	·	Fruit and vegetable products

3.5
Agriculture

(Sources: A.P.E.C.
Canada, Dept. of Agriculture, Economics Branch)

Farm land in the Atlantic Provinces, 1966*

Province	Farm Status (in 000's)				Improved Land (in 000's of acres)			
	No. of non-commercial farms†	No. of commercial farms‡	Total no. of farms	% change in no. of farms from 1961 to 1966	Total area in farms	Areas of improved land	% of improved land	% change in improved land from 1961 to 1966
Nfld.	1.4	0.3	1.7	− 2.5	50	21	42.0	+ 5.0
N.S.	6.8	2.9	9.6	−23.1	1,852	486	26.2	− 2.4
N.B.	5.8	2.9	8.7	−26.1	1,812	639	35.4	−12.9
P.E.I.	3.0	3.3	6.4	−13.1	927	570	61.5	− 1.7
Atlantic Prov.	17.0	9.4	26.4	−20.9	4,641	1,716	37.0	− 6.3

Source: DBS, *Census of Canada, 1966.*
*Statistics for Figure 3.5.
†A non-commercial farm in 1966, one having sales of farm products of less than $2500.
‡A commercial farm in 1966, one having annual sales of products of $2500 or more.

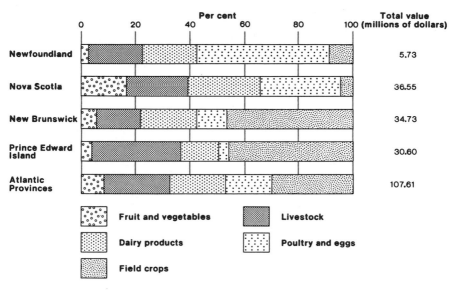

3.6

Value of Commercial Farm Products 1965-1966

decline of 6.7 per cent for Newfoundland, the lowest in the region, emphasizes the importance of subsistence and part-time cash income farming in the Newfoundland-outport economy. Undoubtedly attractive employment in other services and industries and the deterioration of the soil-base has contributed to a faster decline taking place in the Maritime Provinces.

The changes in commercial and non-commercial farming emphasize, in a marked degree, the changes taking place in Atlantic agriculture. (Table 3.11). The actual farm acreage in non-commercial or subsistence agriculture in 1966 was about 2.3 million acres or 49.4 per cent of all farm land. In the period from 1961 to 1966 the acreage of non-commercial agriculture had declined at a phenomenal rate, from a low of 24.3 per cent in Newfoundland to a high of 33.4 per cent in Prince Edward Island. In effect, during the period from 1961 to 1966, the amount of land devoted to subsistence agriculture in the Atlantic region has decreased by almost one million acres or about 30.2 per cent.

Simultaneously, with the decline in non-commercial agriculture, a marked growth that amounted to 8.7 per cent was taking place in commercial farming; the increase for Canada was 11.1 per cent. The increase in acreage varied greatly throughout the region from a minimum rate of 2.5

Table 3.9 Status of farming, Atlantic Provinces, 1966
(in 000s of acres)

	Total land area	Area in farms	Farmland as % of all land	Improved land	Improved land as % of all farmland
Newfoundland	91,549	50	0.1	21	42.0
Nova Scotia	13,057	1,852	14.2	486	26.2
New Brunswick	17,814	1,812	10.2	639	35.3
Prince Edward Island	1,398	927	66.3	570	61.5
Total:	123,818	4,641	3.7	1,716	37.0

Table 3.10 Number of farms of the Atlantic Provinces, 1966
(in 000s of farms)

	Non-commercial			Commercial			Total farms		
	1961	1966	Change (%)	1961	1966	Change (%)	1961	1966	Change (%)
Newfoundland	1.5	1.4	− 6.7	0.28	0.3	+ 7.1	1.8	1.7	− 5.6
Nova Scotia	9.5	6.8	−28.4	3.0	2.9	− 3.3	12.5	9.6	−23.2
New Brunswick	8.7	5.8	−33.3	3.1	2.9	− 6.5	11.8	8.7	−26.3
Prince Edward Island	4.4	3.0	−31.8	2.9	3.3	+13.8	7.3	6.4	−12.3
Total:	24.1	17.0	−29.5	9.3	9.4	+ 1.1	33.4	26.4	−21.0

Table 3.11 Farm acreage, Atlantic Provinces, 1966
(in 0,000s of acres)

	Non-commercial			Commercial			Total farmland		
	1961	1966	Change (%)	1961	1966	Change (%)	1961	1966	Change (%)
Newfoundland	3.7	2.8	−24.3	1.8	2.1	+16.7	5.5	4.9	−10.9
Nova Scotia	143.1	103.3	−27.8	79.9	81.9	+2.5	223.0	185.2	−16.9
New Brunswick	133.9	91.4	−31.7	86.1	89.8	+4.3	220.0	181.2	−17.6
Prince Edward Island	47.9	31.9	−33.4	48.1	60.8	+26.4	96.0	92.7	−3.4
Total	328.6	229.4	−30.2	215.9	234.6	+8.7	544.5	464.0	−14.8

per cent in Nova Scotia to a high of 26.4 per cent in Prince Edward Island – more than twice the Canadian level. A comparison of Tables 3.10 and 3.11 shows that the numbers of commercial farms for the region increased by 1.1 per cent, the commercial farm acreage by 8.7 per cent; the most phenomenal growth, both in the number of commercial farms and commercial farm acreage, took place in Prince Edward Island; a marked increase in the size of farms had taken place in the interval from 1961 to 1966.

It could be expected that the significant trend towards commercial agriculture would be matched by an increase in the size of farm units, particularly in Prince Edward Island. This trend, however, is not proceeding evenly throughout the region: the number of farms in the 70–239-acre class in New Brunswick and Nova Scotia declined by 14.2 per cent and 12.7 per cent respectively; but the number of farms exceeding 240 acres increased by 7.5 per cent in New Brunswick to 1494, and by 6.2 per cent in Nova Scotia to 1320 farms. A persistent trend to larger units above the 70-acre level was evident in Newfoundland. More spectacular was the trend in Prince Edward Island agriculture which, although it showed increases in all farm sizes, was much more marked in classes exceeding 180 acres: the number of these farms increased by 38.2 per cent to 1314, whereas, in commercial farms smaller than 180 acres, the rate of growth was only 4.1 per cent. In the same period the Canadian growth in the number of farms exceeding 240 acres was 10.8 per cent; for units less than 240 acres the rate of growth was 1.5 per cent. In effect, the trend to larger units in the 1961–6 period proceeded at a faster rate in Prince Edward Island than in Canada as a whole. The most rapid growth occurred in 180–239-acre and 240–399-acre classes in Prince Edward Island, in the 400–599-acre class in Nova Scotia and New Brunswick, and in the 240–399-acre class for Canada; whereas, the decline in Nova Scotia and New Brunswick occurred in the two combined classes, 130–179- and 180–239-acres. Although small commercial farms (less than 180 acres) in the Atlantic Provinces comprised 45.7 per cent, they represented a much smaller acreage compared with the rapidly growing large farm units of the region. However, in 1966 the ratio of commercial farms less than 180 acres to commercial farms greater than 180 acres was 0.5:1 for New Brunswick, 0.6:1 for Nova Scotia, 1.5:1 for Prince Edward Island, and 10.6:1 for Newfoundland compared with a ratio of about 1.4:1 for central Canada, indicating that Prince Edward Island more closely resembled central Canada in farm size than the other provinces.

Dairying and livestock accounted for 53.3 per cent and field crops for 23.2 per cent of the activity of commercial farms in the Atlantic Region.

Grain farming is unimportant in the regional picture. The trend in dairying and livestock activities has been downward in both Nova Scotia and New Brunswick, with a small increase in field crops. In Prince Edward Island there has been a marked increase in both livestock and field-crop farming, amounting to 39.2 per cent, compared with an increase of 6.7 per cent for Canada. Field crops are of less importance in Nova Scotia, whereas fruit and vegetable growing are much more important than elsewhere in the region. Generally farming in the Atlantic region is based on mixed agriculture. That is, coarse grains and hay crops are grown to support the raising of livestock; field crops such as potatoes, turnips, tobacco, etc., surplus hay, and grain are sold for cash returns.

The total value of agricultural products produced in the Atlantic Provinces in 1965 was $124.2 million; of this amount non-commercial farms produced $6.59 million or 5.3 per cent. The total sale value from commercial farms amounted to about $107.6 million; of which field crops contributed 30.2 per cent or $32.0 million; livestock accounted for 23.4 per cent or $25.2 million; dairy products for 20.6 per cent or $22.2 million; poultry and eggs for 17.2 per cent or $18.6 million; fruits, vegetables and greenhouse products for 8.4 per cent or $9.1 million. Considerable fragmentation in the relative importance of these segments exists in the structure of provincial commercial agriculture (see Figures 3.5 and 3.6).

In the postwar years, maritime agriculture has been faced with a substantial reduction in both the labour force and in agricultural capital; nevertheless, agricultural production has grown. Increases in productivity may be largely due to better farm management, increased specialization such as tobacco growing in Prince Edward Island, and to the increasing size of large commercial farms. Undoubtedly a major limiting factor to growth is the restricted export market for farm products, particularly in the adjacent mega-urban markets of the northeastern United States.

Increasing costs of capital, inflationary pressures, rising taxes, the official tendency to hold down the price of primary products through various forms of farm prices support, and the attractiveness of off-farm jobs are likely to encourage the marginal farmer to leave the land, thus continuing the trend to the formation of larger farm units. Basically this trend, which has accelerated in the past decade, has occurred at the expense of the low-productivity, marginal farm and of the small commercial unit. The increasing number of small commercial farms in Prince Edward Island suggests that the scarcity of good land may be a factor there.

Most marginal, non-commercial farms provide residence and accommodation features for rural people who are not qualified for any other type of economic activity. These people are likely to resist the consolidation of their holdings with a commercially oriented neighbour, particularly when there has been a tendency in recent years towards increasing market value of land. The Census of 1966 reported 9434 commercial-farm operators in 1965, of whom 33.4 per cent reported off-farm work; but of the 16,959 non-commercial farm operators, 56.4 per cent reported off-farm work. Some authorities expect that the problems of inefficient management and inadequate capital can only be met by an incorporated-type of farm operation rather than through a family-type of ownership. It is interesting to note that because of the rapid growth of a high-value cash crop such as tobacco, farmers in Prince Edward Island found little difficulty in obtaining the necessary financing to manage their operations.

Basically, there are several major problems of maritime agriculture: the high percentage of low-productivity marginal farms, the high cost of modernizing the various aspects of a commercial farming operation, excessive distances to the highly competitive central Canada markets, and severe competition in agricultural products from areas adjacent to the great mega-urban markets of the northeastern United States. Given access to these markets, farm consolidation, agricultural specialization, and efficient farm management could develop rapidly. Without access to these markets the trends in the social structure of agriculture are likely to continue as a gradual process, in step with the growth of the traditional markets for maritime agricultural products. However, serious economic conditions in the manufacturing, servicing, and constructional sectors could induce many to retain the haven of the small farm until the economic weather has improved.

Mineral Resources[4]
Few people consider the Atlantic Provinces to be an important source of minerals though in recent years Labrador has become synonomous with vast iron-ore deposits. The Atlantic Provinces, with the exception of Labrador, are part of the Appalachian geological structures and their mineral resources tend to be similar; the older Ordovician structures in New Brunswick, Nova Scotia, and Newfoundland are a basic source of metallic minerals; mineral fuels, non-metallic minerals, and structural materials are largely associated with Mississippian, Pennsylvanian, or younger structures; the vast iron-ore deposits of Labrador are derived from Precambrian structures. There are over 50 mineral producers in the Atlantic region; many mines are small to medium in size with mill capacities varying from

4 The data for the tables in this section are from: DBS Cat. Nos. 26–201, 1967 and 26–203, 1969.

1000 to 2000 tons of ore a day; fewer than a dozen mines have daily capacities of 3000 to 8000 tons. At the summit are the mines based on the vast Labrador iron-ore deposits which, in scale, rank as one of the world's largest ore bodies with reserves currently estimated at 10.8 billion tons. In today's highly competitive market for capital, large ore bodies have a decided advantage over small ones because of their long-term prospects for stable supplies; moreover, the marketing of minerals is highly competitive. As the bulk of minerals is exported to United States and European markets, the mineral markets are particularly sensitive to industrial supply and demand within these areas.

In 1969 the mineral production of the Atlantic Provinces was $392.6 million or about 8.4 per cent of the total Canadian production. In the period from 1950 to 1960 the Atlantic region production in relation to the total Canadian production declined from 9.4 per cent to 6.9 per cent; the increase for the Atlantic Provinces was at a slower rate than for Canada as a whole – 74 per cent compared with 137 per cent. The decade of the 1960s was marked by a decided upward trend: the Atlantic region production rose from 6.9 per cent (1960) to 8.4 per cent (1969), with the Atlantic Provinces' production increasing at a faster rate than Canadian production – 131 per cent compared with 89 per cent.

An examination of the components of mineral production is given in Table 3.12. Metallic minerals valued at $300.5 million in 1969 constituted 76.5 per cent of the total value of a mineral production of $392.6 million in the Atlantic region, followed by non-metallics, 7.5 per cent, and structural materials, 6.7 per cent, compared with Canadian production of 49.5 per cent for metallic minerals, 9.5 per cent for non-metallics and 9.8 per cent for structural materials. When the regional components are compared with the Canadian production of $4,690.6 million, metallic minerals accounted for 12.9 per cent, non-metallics, 7.5 per cent, structural materials, 6.7 per cent and mineral fuels, 1.9 per cent. The region showed a major weakness in mineral fuels: 7.0 per cent compared with 31.1 per cent for Canada; moreover, there is almost a complete lack of mineral fuel production in Newfoundland, while Prince Edward Island's contribution to the mineral economy is limited to constructional material such as sand, gravel, and stone.

Iron ore in 1969 constituted about 45.5 per cent of the value of all mineral production; copper, lead and zinc accounted for 27.8 per cent; coal, 7 per cent; sand, stone, and gravel, 5.0 per cent. In Newfoundland the value of copper, lead, and zinc production, together with silver, an important by-product, was $38.5 million in 1969, an increase of 4 per cent over 1966; however, the value of New Brunswick's production for these min-

Table 3.12 Mineral production of the Atlantic Provinces (estimated) 1969 (in millions of current dollars)

	Metallics	Non-metallics	Mineral fuels	Structural materials	Regional total
Newfoundland	217.8	14.9	—	6.4	239.1
Nova Scotia	1.4	15.4	21.6	15.8	54.2
New Brunswick	81.3	3.2	5.9	7.9	98.3
Prince Edward Island	—	—	—	1.0	1.0
Atlantic region	300.5	33.5	27.5	31.1	392.6
Component as % of regional total	76.5	8.5	7.0	8.0	100
Atlantic components % of Canadian component	12.9	7.5	1.9	6.7	8.4

erals was $80.5 million, an increase of 17 per cent over 1966. The order of importance varies considerably: in Newfoundland, copper production is more than double the value of lead, zinc, and silver; whereas, in New Brunswick, zinc accounts for 60 per cent, lead for 21 per cent, silver for 10 per cent, and copper for 9 per cent. The proportion of these minerals in each province has remained approximately the same from 1966 to 1969. In Nova Scotia a different group of minerals assumes importance: barite, gypsum, salt, and coal – non-metallics and mineral fuels – these four accounted for 68 per cent or $37 million of mineral production; the value of structural materials production, $15.8 million, is greater than that of the other three provinces combined.

A striking feature of the mineral picture has been the decline in the importance of coal: in Nova Scotia, coal declined in value from $42.8 million in 1966 to $21.6 million in 1970 – a decline of almost 50 per cent; coal production declined in New Brunswick from $7.8 million in 1966 to $5.8 million in 1969. The most spectacular feature was the phenomenal increase in the production of iron concentrates, pellets, and shipping ore, from 16.3 million tons in 1967 to 25.6 million in 1970 – an increase of 58 per cent in three years.

Employment in the mining industry of the Atlantic Provinces in 1967 amounted to 15,327, an increase of 8 per cent since 1961, compared with the Canadian rate of about 11 per cent (Table 3.13). A comparison of earnings in the Atlantic provinces with that of Canada as a whole shows that the average income of $5989 is below the Canadian average of $6824. In this respect, per capita income from mining operations is substantially lower for Nova Scotia and New Brunswick; whereas, the Newfoundland income of $7742 is 13.4 per cent higher than the Canadian average. In terms of value of mineral production per employee the Atlantic region is 34.4 per cent below the Canadian average of $42,887. Value of production per employee varies widely from a low of $10,553 for Nova Scotia to a high of $47,211 for Newfoundland – a figure 10 per cent higher than the Canadian figure. The Newfoundland mining situation emphasizes the impact of scale of the Labrador iron-ore mining operations on mineral production.

Often a single or several important minerals have dominated the mineral structure of each province, with the production coming from several mining operations in local areas. Iron has been the only production mineral in the Labrador mining operations; in the Island of Newfoundland, copper, lead, zinc, asbestos, and fluorspar accounted for 83 per cent; in New Brunswick copper, lead, zinc, silver, and coal for 88 per cent; and in Nova Scotia barite, gypsum, salt, and coal for 65 per cent. Rare and valu-

Table 3.13 Mineral production 1967

	Number of employees	Salaries and wages ($000)	Per employee income ($)	Value of mineral production ($000,000)	Employee value of mineral production ($)
Newfoundland	5,503	42,604	7,742	259.8	47,211
Nova Scotia	7,524	38,027	5,054	79.4	10,553
New Brunswick	2,300	11,168	4,856	89.9	39,087
Prince Edward Island	—	—	—	1.7	—
Atlantic Provinces	15,327	91,799	5,989	430.8	28,107

able minerals such as gold, silver, bismuth, cadmium, and others have been by-products of metallic mining operations.

Structural materials, sand, gravel, stone, etc., in the Atlantic region were valued at $31.1 million in 1969; this amount represented a decline of 9.3 per cent from 1966. Newfoundland and Prince Edward Island experienced gains; whereas in Nova Scotia, structural materials declined from $17.4 million to $15.8 million, and in New Brunswick, from $11.0 million to $7.9 million. The demand for structural materials is related to the domestic market; its trends are closely influenced by construction activity.

A selected number of mining operations in the Atlantic region are discussed to illustrate the concept of scale. Perhaps no mining operation has caught the public imagination more than the Labrador iron-ore mining operations. In terms of the size of deposits, the extent of capitalization, the scale of ancillary services, the fusion of a management organization, and the volume of production from open-pit workings, the Labrador mining operations can best be described as a tremendous undertaking (Figures 3.7 and 3.8).

The metamorphic hematitic iron-ore formations of the Labrador geosyncline contain the largest reserves of iron ore on the continent. The iron-bearing formation in the Schefferville–Knob Lake area consists of hematite–goethite–limonite containing from 45 to 60 per cent iron. The iron content of the deposits at Labrador City, Wabush, and Julian Lake varies from 30 to 40 per cent in specular hematite. Production content of the region may change considerably from year to year depending on the zones that are being mined; the Schefferville product is marketed as direct shipping ore (53%). Through beneficiation the iron content of the Labrador mines may be raised to 60 or 66 per cent.

Reserves of iron ore in the Schefferville area are estimated at 770 million tons; when the potential ore is added, the total iron ore resources amount to 26.32 billion tons. In the Carol Lake–Wabush Lake area iron ore reserves are estimated at 10.05 billion tons, and potential ore at 2.85 billion tons giving total iron ore resources of 12.9 billion tons (Figures 3.7 and 3.8).

To bring Schefferville–Knob Lake deposits into production a consortium of 10 companies – 2 Canadian and 8 American – formed the Iron Ore Company of Canada to organize financing, technical skills, production, and marketing. The 360-mile Schefferville–Sept-Îles railway was completed in 1954; the first ore shipments began immediately. Currently five open-pit mines are in operation – four in Quebec and one in Labrador – producing, in 1968, about 6.5 million tons. In 1962 the IOCC opened the Carol Lake mines in the southwest corner of Labrador; the Carol Pellet

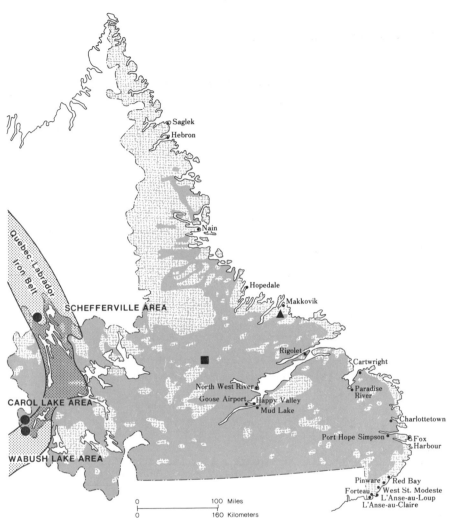

Forestry 1963

Forested area-95% softwood (includes unproductive woodland and burnt-over areas)
Estimated volume of timber: 6.2 billion cubic feet of material 4-9" DBH
1.2 billion cubic feet of material 10" and over DBH

Bog and barren

Minerals

● Iron mining areas
Occurrences of potential importance
▲ Uranium (U)
■ Uranium (U), Niobium (Nb) and Beryllium (Be)

3.7

Mining and Forestry in Labrador

(Sources: Newfoundland Dept. of Mines, Agriculture and Resources
                 Canada   D.E.M.R., Mineral Resources Division)

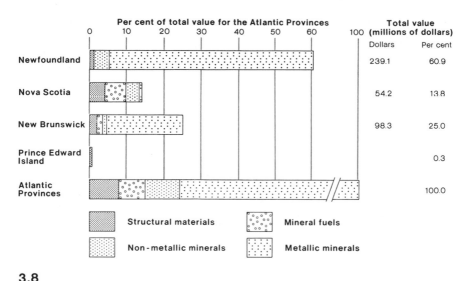

3.8

Mineral Production in the Atlantic Provinces 1969

(Sources: D.B.S. Mineral Industries 1969 and Mineral Production 1969)

Iron ore resources of western Labrador, 1969*
(in billions of metric tons)

	Reserves†	Potential	Total resources‡
Schefferville area	0.77	25.55	26.32
Carol–Wabush L. area	10.05	2.85	12.90
Total	10.82	28.40	39.22

Source: Gross, G.A., 'Iron Ore Deposits of Canada and the West Indies,'
in Survey of World Iron Ore Resources, United Nations, 1970.
*Statistics for Figure 3.7.
†Reserves refer to measured, indicated and inferred ore.
‡Resources include reserves and potential ores.

Company, composed of a consortium of seven American companies, processes IOCC hematite and magnetite concentrates for shipment. In 1964
production at Carol Lake reached 6.3 million tons of iron concentrate and
pellets. In 1965 Wabush Mines, a new consortium formed by two Canadian and six foreign companies, mainly American, opened the iron deposits at Wabush Lake; by 1968 production had reached 6 million tons of

iron concentrate and pellets. Canadian Javelin, currently seeking financing and markets, expects its deposits to produce about 12 million tons of iron concentrate annually.

In 1969 production of iron ore from the Newfoundland–Labrador mining operations reached about 15 million tons valued at about $179 million. Crude ore production in the region, from open-pit, reached about 49 million tons in 1970: of this amount, 24.4 million tons, yielding 38 per cent iron, came from the Carol Lake operations; 16.6 million tons from Wabush Mines, yielding 31 per cent iron, and 7.9 million tons, yielding 54 per cent, from Quebec–Labrador mines at Schefferville. The crude ore yielded a total of 24.4 million tons of usable iron ore, having a concentration of 59 to 66 per cent. Of the concentrates shipped through Sept-Îles, about 18 per cent entered the domestic market, 47 per cent was exported to the United States, and about 36 per cent to west European markets. A large part of producton was shipped to owner-companies in Canada and the United States. The Iron Ore Company of Canada expects its annual production capacity from its Quebec–Labrador operations to reach 30 million tons of concentrates by 1972.

To accommodate the increased volume of shipments, dock and port facilities at Sept-Îles for servicing 200,000-ton bulk carriers, were completed in 1969; the iron-ore loading dock, costing about $13 million, is one of the world's largest. The completion of the Seaway in the mid-1950s permitted bulk shipments of ore to American and Canadian markets on the St Lawrence River and the lower Great Lakes. In the late 1950s the development of power for the west Labrador mining operations began at Twin Falls on the Unknown River; by 1967 over 300,000 H.P. was developed. This undertaking represented an investment of about $60 million.

The Iron Ore Company of Canada has estimated that it and its subsidiaries will have spent almost $1 billion by 1971 to bring its mines in the Quebec–Labrador area up to the 1971-level of production. The IOCC in 1970 began the expansion of its 12 million-ton-a-year concentrator to 22 million tons capacity at Labrador City; a 6 million ton-a-year concentrator is to be constructed at Sept-Îles to upgrade the Schefferville ore; both are to be completed in 1973. The Labrador City expansion is to cost about $150 million and the Sept-Îles expansion about $140 million. The construction of the two projects is expected to provide 3000 construction jobs and 800 permanent jobs. The mining operations at Labrador City and Wabush have provided the basis for two urban communities: Labrador City at Carol Lake, begun in 1960, had reached a population of about 9500 in 1971; Wabush had grown from its 'founding' in 1965 to a population of about 3300 in 1971.

Copper mining is the oldest important mineral industry in Newfoundland. By the last quarter of the nineteenth century, Newfoundland had gained world prominence in copper production. The industry, centred around the numerous small mines of high grade ore, was located in the mineralized belt around Notre Dame Bay. Most mines were shallow operations and were quickly worked out; however, some of the larger mines continued to be worked until World War I. Mineral exploration in the area in the 1960s increased the number of producing mines to six. Most of these mines were underground workings, small to medium in size with copper grading from 0.84 to 0.99 per cent; the mines were of 1200 to 2000 tons per day capacity. Copper concentrates were shipped to the Gaspé Copper Mines smelter at Murdochville, Quebec. As exploration failed to locate extensions of the orebodies, there has been a succession of mine closures: Tilt Cove closed in 1967, Little Bay in 1969, Gullbridge in 1971, and Whaleback is to close in mid-1972. The American Smelter and Refining Company mines at Buchans, in the southeast area of the Notre Dame Bay mineralization belt, is the sole lead producer in Newfoundland. Operations began in 1927. In 1970 its ore graded as follows: lead 7.02 per cent, zinc 12.68 per cent, copper 1.11 per cent, and silver 3.73 oz. The mill capacity is 1250 short tons per day; proven and probable ore reserves are sufficient to support mine production for another 8 to 10 years. The various mineral concentrates from Buchans are shipped to west European countries and to the United States.

In the industrial-mineral field, asbestos fibre from Baie Verte has been recovered from open-pit operations since 1963; mill capacity is 5000 tons a day. A fluorspar mining operation began at St Lawrence in the Burin Peninsula in 1942; mill capacity is 900 tons a day; the concentrate is shipped to the aluminum operations at Arvida, Quebec. In 1968, an open-pit mining operation for silica began at Ville-Maria; mill capacity is 2200 tons a day. The silica is shipped to Long Harbour, Placentia Bay for use in manufacturing elemental phosphorus.

Ordovician rocks, principally volcanics, are the hosts for the massive pyritic-base metal deposits of the Bathurst area of New Brunswick. In 1952 a major base-metals deposit was discovered southwest of Bathurst; a staking rush followed that was without precedent in New Brunswick; some 20 deposits have been discovered. New Brunswick has come to be recognized as one of the major lead–zinc districts of the world. The original discovery, No. 6 Mine of Brunswick Mining and Smelting, was brought into production in 1966, although its No. 12 Mine came into production in 1964. Heath Steele Mines commenced production in 1957, suspended operations in 1958, to recommence in 1962. Nigadoo River Mines commenced production in 1967 and Caribou (Anaconda) in 1970. BM & S Corporation's smelter began operation near Bathurst in 1966, producing refined lead and zinc. Brunswick Mining and Smelting has become Canada's second producer of primary lead and the fourth of zinc. A fertilizer plant at Belledune was established for the production of diammonium phosphate. The smelting and fertilizer operations are subsidiaries of BM & S mining operations. BM & S is composed of a consortium of six companies, in which Noranda Mines Ltd. acts as manager and supervisor, and provides technical assistance to BM & S and its subsidiaries; in effect, BM & S has become part of the Noranda group of companies.

Brunswick Mining and Smelting is the largest mining operation in New Brunswick. In 1970 the ore reserves from its two deposits were estimated to contain about 97 million tons of ore. The production of ore in 1970 amounted to 2.5 million tons or about 8300 tons per working day. The grade of the ore varied from 5.86 to 7.54 per cent zinc, 2.12 to 2.93 per cent lead, 0.32 to 0.33 per cent copper, and 1.84 to 2.19 ozs. silver. Production of lead–zinc concentrate amounted to 188,000 tons, copper concentrate about 10,200 tons, and over 1 million ozs. of silver. In addition, valuable amounts of cadmium, antimony, and bismuth were recovered.

Ore is recovered from both open-pit and underground workings. Substantial improvement in smelter operations for the recovery of metals and increased production at both mines is expected to raise production to over 9000 tons a day; by 1971 the grade is expected to average 12.3 per cent, combined lead and zinc, and yield a 65 per cent concentrate. The fertilizer operation in 1970 was operating at 29 per cent plant capacity.

The four mines: Brunswick, Heath Steele, Nigadoo, and Caribou, operating in the Bathurst mineralized belt, employed a staff of some 2800 in 1971 – a number greater than the total engaged in the New Brunswick mining operations in 1967; moreover, BM & S employees numbered 1970 in 1971 or 85 per cent of the provincial figure in 1967.

The most critical situation in terms of Nova Scotia's mineral economy is related to coal and steel. In 1967 the Cape Breton Development Corporation (DEVCO) acquired the coal interests of the Dominion Steel and Coal Corporation, and anticipated a 15-year phasing out of coal-mining operations. In 1968 the Sydney steel plant was taken over by the Nova Scotia government from Hawker Siddeley Canada Ltd. and operated by DEVCO with federal government participation.

The high volatile bituminous coal deposits of Nova Scotia, produced mainly from underground mines at Sydney, are of Carboniferous age. Although coal production has been declining steadily to almost 2.1 million tons in 1970, valued at $22.2 million, Nova Scotia produced about 24 per cent of Canada's output of bituminous coal; Nova Scotia's output was exceeded only by Alberta and British Columbia. DEVCO is the largest producer; its production from four collieries on Cape Breton Island and a mainland plant at Thorburn, N.S., amounted to 1.98 million tons or about 93 per cent of the Province's coal production. The DEVCO operation employs about 3541 persons or 96 per cent of those employed in the coal-mining industry and about one-half of the employees in the entire mining industry of the province. Output per man-day in 1970 was 2.5 short tons compared with 1.7 for New Brunswick, 7.9 for Alberta, and 4.9 for Canada; moreover, the value of production per employee for Nova Scotia was approximately $6080.

The chief markets are domestic, industrial, and thermal power generation. The Sydney steel plant consumed about 115 thousand tons of coke, produced from the Corporation's coke-oven plant at Glace Bay, and about 28,000 tons of coal for raising steam. About 69 per cent of the province's production is consumed within the province, and 26 per cent is consumed in Ontario and Quebec.

DEVCO has been modernizing its existing collieries while continuing with the development of a new mine at Lingan that is scheduled for opening in 1974. The reserves of coal – measured, indicated, and inferred – in Nova Scotia amount to 4.1 billion tons or 3.5 per cent of the total Canadian coal estimate. The most important reserves in the Sydney area lie in submarine areas off the coast. In 1946, the Sydney coal field reserves were estimated to contain about 2.34 billion short tons, of which 976.8 million were available for immediate production, a further 457.3 million tons were considered recoverable, and the remaining 911.0 million tons were assumed to exist. The estimate is based on seams not less than 2 feet thick at a vertical depth of not more than 4000 feet.

Marketing prospects for mineral products vary considerably from season to season and from year to year; such trends closely reflect consumer demand for metal products. In other areas new technology has brought changes that are reflected in decreased consumption of coal; at the same time, the tempo in off-shore oil-drilling has increased, with the most promising discovery that made on Sable Island in 1971. Commercial production of crude would reduce crude imports in the Atlantic region; important benefits should be reduced fuel costs in oil-fired thermal generating plants and for domestic, commercial, and industrial uses.

Recent developments in the demand for coal are promising. The production of coal in 1970 in western Canada increased by 90 per cent over 1969, and in the same period in eastern Canada decreased by 24 per cent. The growth of western coal production was directly related to Japanese requirements and to the domestic thermal electric generating industry. Coal production is expected to increase in both Nova Scotia and New Brunswick in order to meet the requirements of the electric generating utilities of the Atlantic Provinces. DEVCO is expected to use an increasing proportion of its coal production for the making of metallurgical coke. Further increases depend on continuing worldwide shortages, particularly of high quality coking coal.

Among the metallic minerals, the consumption of silver, particularly for industrial and commercial purposes, has been increasing more rapidly than production. Silver supplies are inelastic – all of the Atlantic Provinces production and about 85 per cent of Canadian production are derived from the production of copper, lead, and zinc ores. In the world copper markets, about 93 per cent of the copper consumed in the non-Communist world is used in the United States, Western Europe, Japan, and Canada. Thus, the single most important factor in the world copper market is the condition of the United States economy – any fall in prices would reduce tonnages, forcing smaller and marginal mines to close. Reduced production from Chile, Peru, and Iran should strengthen financing prospects for new mining production in the near future. Sharply rising costs, however, require a copper price of over 50¢ a pound to justify investment in a typical North American capital-intensive mining operation. Copper consumption has been growing at about 4.5 per cent a year so that increases in mine production would not likely exceed consumption. Because of the large amounts of capital involved, few operators would risk new financing in a weak market. In general, the same situation applies to zinc, lead, and iron; oversupply and large inventories tend to force prices down and to reduce production in the mines. Certain important constraints on mineral processing which affect policy-making decisions are imposed by refineries and smelters located in the importing industrial nations. In western Europe, long-established capacities have been developed and sustained largely on imported concentrates in countries often with little or no mineral resources of their own. Custom refineries and smelters solicit concentrates; however, to develop a producing mine requires that contracts be tied to existing smelters. Smelters and refineries attract railway and port facilities, capital and financial investments, labour and skills that cannot be abandoned but must be retained or expanded. Tariffs are imposed to protect this structure and mineral demands are often modified in the desire of industrial nations to restrict the imports of refined metals in favour of mineral concentrates and raw material. Although high internal metal prices in a processing country may allow imports, this policy has a twofold effect, namely, to protect the processing country's capital investment and to discourage the establishment of processing facilities in the product's country of origin. Again, industry is closely integrated in industrial nations; thus, a by-product such as sulphuric acid may be used in local chemical industries, but in Canada it could present a disposal problem. In effect, these barriers prohibit a greater degree of pre-export processing from being done in Canada; it is a difficult task for Canada, one of the world's largest mineral producers and exporters, to sell metal; it is not a problem to dispose of concentrates. Most constraints are beyond the control of the exporting country. Once interdependent tariff structures relating to ores, concentrates, and metals have become established on a multinational basis, the mineral-flow patterns are exceedingly difficult to change.

A new trend, which would strengthen considerably the position of the importing nations, is developing: industrial nations with growing exchange reserves, who are unwilling to hold either Canadian or American dollars, may stockpile Canadian resource materials. The stockpiling of Canadian copper, lead, and zinc concentrates (with their content of precious metals) in exchange for u.s. dollars is aided by the recent monetary realignment. Thus, because of the rise in value of their currencies against the Canadian dollar, importing industrial nations may buy Canadian metals, ores, and concentrates more cheaply than before realignment. It would seem that the future demand for Canadian mineral resources is strong.

In 1970, the value and percentage of total Canadian metal and mineral exports to the United States were $2,734.4 million or 54.0 per cent; the value and percentage of Canadian exports to the United Kingdom were $720.3 million or 14.2 per cent, and to the European Economic Community $555.0 million or 11.0 per cent. Thus a healthy United States economy would tend to strengthen mineral prices and increase mine production; furthermore, a continuing strengthening of western European economies would provide a broad consumers' base over the long term for a healthy mineral industry both in the Atlantic Provinces and in Canada generally.

Energy

There is a relationship between the rate of increase of energy consumption and the stage of a region's economic development. Energy is closely involved in the economy and an increase in economic output is likely to result in an increase of energy consumption. Increased utilization of electricity has made individual users more dependent on utility systems for essential functions such as heating and lighting and for electric-powered household equipment. In this and in the commercial and industrial sectors a high level of reliability is essential. The demand for energy, moreover, can be expected to rise in the future provided that economic growth and the modernization of our living pattern continues. Because of the substitution of fuels and the changing structure of the economy, it is difficult to relate relative increases in the economy to the consumption of energy. If the substitution of oil and gas for wood and coal has passed its peak of the 1950s, then energy consumption and overall economic growth in the future should tend to move closer together. It is expected that the use of petroleum will continue to grow; coal is expected to level off; the use of wood for fuel should continue its downward trend; the use of hydro-electricity is expected to grow, but, in time, to be out-paced by thermal-generated electricity.

In 1960 about 306×10^{12} Btu's of energy in the form of hydro-electricity, coal, wood, and petroleum fuels were consumed in the Atlantic Provinces: this amount represented an increase of 23 per cent since 1950 or an annual rate of growth of 2 per cent; Canadian consumption in the same period rose by 35 per cent to 3285×10^{12} Btu. The Atlantic Provinces' proportion of the total Canadian consumption of energy fell from about 7 per cent to about 6.4 per cent. Moreover, the per capita use of energy increased from 1950 to 1960 from 1850 kwh to 2600 kwh in the Atlantic Provinces; consumption was about 42 per cent of the Canadian level. The cost of electric power in the Atlantic Provinces is above the Canadian average – for all manufacturing industries in the Atlantic Provinces fuel and electricity represented 9.7 per cent of value added, compared with 5.6 per cent across Canada. Thermal power is expensive because of higher fuel costs in the Atlantic Provinces.

A major change occurred in the 1950–60 period; new fuel sources were substituted for the traditional sources of energy. The share of petroleum fuels has increased from 26 to 59 per cent, hydro-electricity from 2.5 to 3.1 per cent, whereas coal consumption has declined from 57 to 29 per cent and wood from 15 to 9 per cent. In the use of fuel for household heating in the Atlantic Provinces, 32 per cent used wood as their principal fuel, 48 per cent used oil, and only 20 per cent used coal or coke; in New Brunswick and Newfoundland wood-consuming households exceeded coal-burning households by a wide margin. The price of coal has increased considerably in comparison with fuel oil and was a major factor in the substitution of oil for coal, particularly in the thermal generation of electricity. In both Prince Edward Island and Newfoundland, oil is almost exclusively used for fuel in thermal-generating stations; in Nova Scotia over 80 per cent of the generating fuel is coal, and in New Brunswick less than 40 per cent of the thermal stations are coal-fired.

The most rapid advances in the use of electrical energy have occurred in commercial and domestic uses which reflect the importance of the service sector in the post-industrial economy, and the use of transfer payments which stabilize domestic consumption. Thus the current growth in the consumption of electrical energy continues to outgrow advances of economic output or population growth. The use of fuel oil for the generation of thermal electricity in the Atlantic Provinces is expected to exceed its use for shipping in the 1970s, whereas the use of coal for thermal generation is likely to equal all other uses in the early 1970s.

The generation of electrical power varies considerably throughout the provinces. In the past, energy requirements for industrial or community uses were achieved through the conversion of steam power to electricity or, in recent years, through the use of electric-diesel units. These local systems were privately owned and highly fragmented: each served local markets that were served by and dependent on a single supplier of power. The integration of independent power sources has provided systems with extensive and interconnecting transmission systems within provincial boundaries, thus assuring a continuity of power from a diversity of power sources (Figure 3.9). The main systems of Nova Scotia and New Brunswick operate as the Maritime Power Pool; moreover, these expanding systems include ties across the New Brunswick border with the United States. Integrated systems have provided economies of scale in the distribution and sale of energy: there has been marked improvement in the security of power supply and a high level of reliability; moreover the interconnected power grid is the surest form of security and reliability. Other advantages of scale lie in the opportunities to install and use large generating facilities; besides, the characteristics of an interconnected system tend to be complementary. In the present growth and development of interconnected systems, interregional ties are being extended as economic and technical conditions permit. Thus Churchill Falls power is being transmitted to Montreal, with a smaller part of the ultimate output eventually going to the Island of Newfoundland.

Both the manufacturing and utility aspects of the industry have taken advantage of economies of scale through network integration to provide more efficient distribution of energy, through improved design and operating performance of installation. In order to hold costs down in the generation of energy the trend is towards larger units of 150 megawatts or higher. In the acquisition of capital for energy distribution the major elements to be considered for electrical-utility systems are generation 48 per cent, distribution 25 per cent, transmission 21 per cent, and other 6 per cent. Generally the most favourable hyro-electric sites have been developed and increasing environmental restrictions on thermal-plant siting

Transmission lines 1969

Existing	Under construction	Kilovolts
———	- - - - -	66-199
———	━━━━	200-299
	- - - - -	300-399
	━━━━	400 and over

Generating stations 1969

Existing	Under construction (including extension to existing plant capacity)	
●	○	Hydro-electric
▲	△	Thermal-electric

Only stations with total installed generating capacities of not less than 1500 kilowatts are shown

3.9

Electrical Energy

(Source: Electric Power in Canada 1969)

Installed generating capacity in the Atlantic Provinces*
(in 000's of kilowatts)

	Installed capacity 1967				Installed capacity 1969				Annual % increase in all power
	Hydro	Thermal	Total power	Thermal as % of total	Hydro	Thermal	Total power	Thermal as % of total	
Newfoundland	696	107	803	13.3	820	128	948	13.5	9.0
Nova Scotia	152	544	696	78.2	163	773	936	82.6	17.3
Prince Edward Island	—	57	57	100.0	—	77	77	100.0	17.5
New Brunswick	262	532	794	67.0	563	652	1215	53.7	26.5
Atlantic Provinces	1110	1240	2350	52.8	1546	1630	3176	51.3	17.6

*Statistics for Figure 3.9.

have tended to force increased capital investment in transmission systems that are regionally interconnected.

The generation of power varies considerably throughout the Atlantic region (Table 3.14); the percentage of installed hydro-generating capacity electricity in contrast to thermal-generated electricity in 1969 ranges from nil in Prince Edward Island to 86.5 per cent in Newfoundland; hydro-generated power accounts for 17.4 per cent in Nova Scotia and 46.3 per cent in New Brunswick. Slightly over half of the entire power generated comes from thermal plants. The annual rate of increase in total installed capacity of electricity from 1967 to 1969 ranged from 9 per cent in Newfoundland to 26.5 per cent in New Brunswick. The average 17.6 per cent rate of increase for this period is greater than the Canadian average of 10.4 per cent. Thermal power has been increasing at a greater rate than hydro; the decline in the relative position of thermal power in New Brunswick rose from the addition of 300,000 kilowatts of Mactaquac power in 1968.

Provincial Commissions have emerged as the major suppliers of electrical energy. The Nova Scotia Power Commission supplies about 80 per cent of current provincial requirements. The New Brunswick Power Commission supplies 75 per cent of the New Brunswick total. Two companies provide Prince Edward Island with thermal electricity. Of the 946.9 thousand kw produced in Newfoundland about 25 per cent is in Labrador and is largely controlled by the mining corporations. In the Island of Newfoundland the Newfoundland and Labrador Power Comission controls 75 per cent of the total generating capacity of 698.4 thousand kw on the Island. As elsewhere in the region, mining and pulp and paper companies control most of the remainder. Generally there is an increasing trend for smaller mining operations to purchase power from utilities, thereby taking advantage of larger unit sizes and operational flexibility offered by interconnected systems.

The size of power generation units varies considerably throughout the Atlantic Provinces. The total capacity of thermal-generating stations in Newfoundland varies from 2500 to 30,000 kw; whereas, three hydro-power developments have a total capacity of about 659 thousand kw or about 70 per cent of the province's entire capacity. In New Brunswick two hydro stations (412,500 kw) and two thermal stations (383,365 kw) combined, represent about 66 per cent of New Brunswick total generating capacity. In Nova Scotia some 29 hydro-power developments generate a total capacity ranging from 2000 to 18,000 kw yielding a total 163,000 kw; whereas four thermal stations (583,000 kw) generate slightly over 62 per cent of Nova Scotia's total generating capacity.

Table 3.14 Installed generating capacity in the Atlantic Provinces (in 000s of kilowatts)

	Installed capacity 1967				Installed capacity 1969				Annual % increase in all power
	Hydro	Thermal	Total power	Thermal as % of total	Hydro	Thermal	Total power	Thermal as % of total	
Newfoundland	696	107	803	13.3	820	128	948	13.5	9.0
Nova Scotia	152	544	696	78.2	163	773	936	82.6	17.3
Prince Edward Island	—	557	57	100.0	—	77	77	100.0	17.5
New Brunswick	262	32	794	67.0	563	652	1215	53.7	26.5
Atlantic Provinces	1110	1240	2350	52.8	1546	1630	3176	51.3	17.6

The years 1969 and 1970 in the Atlantic region were active either in terms of transmission or generation expansion. In New Brunswick power development, the installation of the 100 MW (Megawatts) thermal generating unit at Dalhousie was the principal addition to plant capacity in 1969; in 1970, a fourth 100 MW unit was being prepared for the Mactaquac hydro station – two further units of the same size are to be added later. Most effort in 1970 was directed towards transmission development: the most significant was the construction of the 345 kV transmission interconnection with utilities in New England, providing the first major inter-regional tie between the Maritime region of Canada and the United States. Planned for service in 1972 is the 320 MW capacity, high-voltage direct current system converter terminal in New Brunswick. This system will permit the purchase of substantial quantities of surplus Churchill Falls power from Quebec Hydro; later it will allow economic interchange of power to the benefit of both provinces.

Nova Scotia in 1969 added thermal capacity totalling 230 MW to plants at Trenton and Point Tupper. Future additions at Point Tupper will make the Point Tupper station the largest in the province. Currently Tuft's Cove station is being expanded by a 105 MW generating unit. Transmission capacity was increased by 88 miles of 138 kV circuit in 1969, and by 152 circuit miles of 69 kV, 138 kV, and 230 kV systems in 1970.

Prince Edward Island's generating capacity was increased by 18 per cent with the construction of a 14 MW gas turbine generating unit at Borden in 1970.

The year 1970 was one of extraordinary expansion for the power industry in Newfoundland, with a 30 per cent increase in installed generating capacity: the installation of two 76.5 MW units of the 459 MW Bay d'Espoir hydro-electric station and the installation of a 150 MW generating unit at the Holyrood thermal station were the basis for the expansion. A second 150 MW generating unit is to be installed at Holyrood in 1971. Future generating expansion in the Island of Newfoundland will probably take place on the western part of the Island. The province's transmission network was extended substantially: in 1969 transmission capacity was increased by 55 miles of 230 kV line; in 1970 a second 230 kV circuit from Holyrood to St John's was completed; moreover, a second 230 kV circuit to Newfoundland's west coast is to be built shortly. In addition, 150 circuit miles of 69 kV line were constructed to provide connection to areas previously served by local diesel-generating units. Thus as the transmission network is expanded, the isolated diesel units of Newfoundland's outport communities will be phased out.

In the history of Canadian hydro-power development, there is none that approaches in magnitude the Churchill Falls undertaking in Labrador. By 1976 the installed capacity of 11 hydro-generating units, rated at 5,225,-000 kilowatts or 7,000,000 horsepower, will yield an annual production of 34.5 billion kilowatt hours; the plant's capacity will equal 13 per cent of all Canadian thermal and hydro power capacity in 1969; it will be 66 per cent greater than all the installed capacity of the Atlantic Provinces for 1969. The first regular power was delivered in December 1971 to Quebec Hydro with the installation of the first two turbines and the completion of the first of the three interconnected 735 kV transmission systems; commercial deliveries to Quebec Hydro are planned for May 1972.

In 1953, a consortium of five companies, mainly British, formed the British Newfoundland Corporation Ltd. (BRINCO) to develop hydraulic and other resources in Newfoundland and Labrador. When in 1966 a letter of intent to purchase most of the power was signed with Quebec Hydro, BRINCO formed the subsidiary, Churchill Falls (Labrador) Corporation Ltd., to develop the Churchill Falls power project; it is composed of a consortium consisting of Hydro Quebec, BRINCO, Province of Newfoundland and Labrador; Wabush Mines and Iron Ore Company of Canada hold equity interests. BRINCO was granted exclusive mineral rights for a 20-year period over more than 50,000 square miles of Newfoundland–Labrador; it obtained rights to develop river systems in Labrador and the Island of Newfoundland. BRINCO thus was allied with strong financial and corporate management and the principal purchaser of its power, Hydro Quebec; construction began in 1966.

Although cost estimates established that $946 million would be required to complete the project, about $1073 million were available. Employment at the site exceeded 6000 in 1970, the peak year; off-site employment in the turbine and generators' fabrication required the equivalent of 600 employees over the 6-year period. At the end of construction in 1974, Churchill Falls is expected to have a permanent population of about 1000.

Water for the hydro-power project is collected from the 26,700 square miles of the Upper Churchill basin and stored in the primary reservoir above Churchill Falls. Rising waters in the reservoir have overflowed Lobstick, Michikamau, and local lakes forming a water body with an area of some 3000 square miles; the reservoir is expected to provide a regulated flow of 49,000 cubic feet of water per second. The surface water level at the intake is 1471 feet above sea-level; the turbines are installed at an elevation of 385–419 feet, and as the penstocks are inclined at about 30 degrees from the vertical a gross operating head of 1060 feet is provided. Each of the 11 penstocks is 1400 feet long and 20 feet in diameter and discharges the waters into its turbine chamber; each hydraulic turbine is rated at 648,000 horsepower. The excavation of the power house was completed in 1970; it is enclosed in underground rock and measures 972′ × 81′ × 154′, and is considered the world's largest underground power house. Two tailrace tunnels, each 45′ × 60′ × 5550 feet long, discharge the flow into the lower Churchill River.

Power will be transmitted over a distance of 126 miles to the Quebec–Labrador boundary for delivery to Hydro Quebec. The development of Churchill Falls is considered the largest single-site, hydro-electric project in the world; moreover, there remains an additional 3,000,000 horsepower to be developed on the lower Churchill River.

Considerable attention has been given in recent years to the development of power from the Bay of Fundy tides in favourable estuaries that have a potential of many billions of kwh annually. An appraisal of economic power potential by the federal government and by the governments of Nova Scotia and New Brunswick indicated that a site at the entrance to Cobequid Bay (Economy Point to Cape Tenny) at the head of Minas Basin would provide lower cost energy than sites elsewhere in estuaries of the Bay of Fundy, St Mary's Bay, or the Annapolis Basin. The magnitude of the tide increases as the tides move up the bay; the spring tide reaches 53 feet in Cobequid Bay; however, the average tide moves about 105 billion cubic feet of water into Cobequid basin; thus, each 24 hours and 50 minutes this volume of water is moved twice into the Bay and twice out of it.

The Fundy tides are regular; variations in the ranges within each lunar month are comparatively small; the head and flow for available use can be predicted for many years; and, moreover, a tidal power plant is not a source of water or atmospheric pollution. The major problem is matching the timing of tidal movements to the timing of consumer demand for power; it is, however, a renewable resource, whereas thermal plants depend on a fuel which, once used, cannot be replaced.

Studies indicated that the Cobequid Bay site would provide greater energy production and firm peak capability, and that the unit cost of power would be lower than at other sites. However, the output from the site could be much in excess of the New Brunswick–Nova Scotia power requirements for a considerable period in the future. Because of production costs, together with transmission costs, the Fundy power would not be competitive with power produced by New England utility systems. Nevertheless, prediction of power requirements to 1990 indicate the need for a tremendous increase in generating capacity for northeastern North America: from an annual consumption of 6445 million kwh in 1968 to a projected consumption of 31,000 million kwh for the Maritime Provinces by 1990; and, from an annual consumption of 41,782 million kwh in 1965 to a projected consumption of 210,000 million kwh for the New England states by 1990.

A single-effect optimum-sized development, designed to generate power only, and in one direction, usually when the basin is discharging on the ebb tide, requires capital expenditure of $474 million to produce 6500 million kwh annually; that is, at a production cost of about 6.5 mills per kwh. A double-effect optimum-size development is designed to generate energy on both the flood and the ebb tides, and thus to convert tidal energy into firm capacity. The installation could provide 1500 mw of dependable capacity for 3 months, for example, and, for the balance of the year, 7560 million kwh. The capital required for this installation would be about $724 million; a gas-thermal plant would provide the same capacity for a capital cost of $150 million; the cost per kilowatt per year of tidal power would be about $37 constrasted with a gas-turbine thermal plant of less than $10. During the 9-month period when energy is the only output, power generation must follow the tidal cycle, with the tidal plant generating close to installed capacity. Since existing large thermal plants convert fuel to electricity at almost 3 mills per kwh, in the present state of market demand and technological development, tidal power is not competitive with thermal power. On this basis, and at the time, because of the necessity for an extensive capital investment and the long construction period, the development of power from the Fundy tides was not considered feasible.

4 Cities: Function, Form and Future

C.N. FORWARD

The urban system of the Atlantic Provinces is a poorly articulated satellite of the Central Canada system centred on Montreal and Toronto. Halifax is the largest city, but it does not function as a regional primate. Fundamentally, the system is comprised of three tenuously linked provincial hierarchies, each maintaining strong ties with the national metropolises. Though key centres of British North America before Confederation, Halifax, Saint John, and St. John's eddied in economic backwaters while witnessing the meteoric growth of inland cities during the first century of nationhood. They were swept into the orbit of Montreal and Toronto, none of them able to assert a convincing regional dominance.

The geographical, political, and social fragmentation of the Atlantic Provinces region sets it apart as unique among the major regions of Canada. The islands and peninsulas are like pieces from different jig-saw puzzles that do not fit together. Transportation systems have been ineffective in drawing the sub-regions together to focus on one major urban centre. Hence, a number of separate hinterland areas that are tributary to several urban centres of similar size have developed independently. Historic factors during the long period of settlement also have contributed to the fragmented pattern that exists. Independent political units were established at an early date and economic development proceeded along separate, but sometimes parallel, lines. Having been colonized by different groups, ethnically or socially, each sub-region developed a local identity and point of view through a long period of time. Benefits have accrued, however, in that even the largest cities remain liveable, distinctive places built on a human scale and etched with lines of character.

All of the urban centres recording 10,000 population or over in the 1966 census are considered in this chapter, but the major emphasis is placed on the few at the top of the urban hierarchy, those in the 100,000 or larger range. The population figures refer to metropolitan areas or major urban areas, wherever these exist. Halifax, at about 200,000, is twice the size of the next largest cities, Sydney, Saint John and St John's,

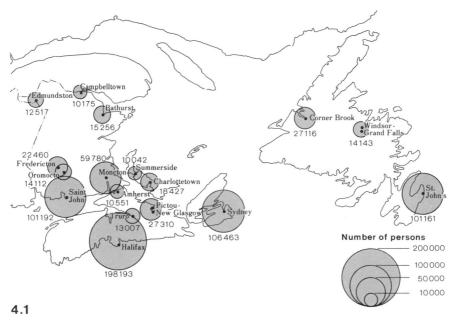

4.1

Urban Centres in the Atlantic Provinces 1966

which form a trio in the 100,000 category (Table 4.1 and Figure 4.1).
Sydney differs from the rest in being a truly 'dispersed' city with a number
of distinct nodes scattered over a large area. Moncton stands alone at
about 60,000, followed by a group of four in the 18,000 to 27,000
bracket, and finally, eight other cities ranging from 10,000 to 15,000
population. Another 'dispersed' city is centred on New Glasgow, including
Pictou, with a total population of 27,000. The lack of a major metropolis
to act as a focus for regional development is a weakness of the urban
system.

In this regard it is of interest to compare the urban hierarchy of the
Atlantic Provinces with that of British Columbia. The west coast province
is a large, compact block of territory with a population of about two
million people, approximately equal to that of the Atlantic Provinces.
Settlement of both east and west coast regions occurred through an approach from the sea and the port function was a key factor in urban growth
on both coasts. One great difference was that Vancouver became the
dominant port and railway terminus on the west coast, while Halifax
shared this function with Saint John in the Maritimes and with St
Lawrence River ports farther inland. The urban hierarchy that evolved in

Table 4.1 Urban population 1871–1966[1]

	1871	1881	1891	1901	1911	1921	1931	1941	1951	1961	1966
New Brunswick											
Bathurst	—	—	—	—	—	3,327	3,300	3,554	4,453	5,494	15,256
Campbellton	—	—	1,782	2,652	3,817	5,570	6,505	6,748	7,754	9,873	10,175
Edmundston	—	—	—	—	1,821	4,035	6,430	7,096	10,753	12,791	12,517
Fredericton	6,006	6,218	6,502	7,117	7,208	8,114	8,830	10,062	16,018	19,683	22,460
Moncton	1,650	6,622	10,462	10,876	13,395	19,538	23,139	25,963	39,624	55,768	59,780
Oromocto	—	—	—	—	—	—	—	—	—	12,170	14,112
Saint John	52,273	52,050	48,681	51,213	53,617	61,218	62,739	70,927	78,337	95,563	101,192
Prince Edward Island											
Charlottetown	7,872	10,345	10,098	10,718	9,883	10,814	12,361	14,821	15,887	18,318	18,427
Summerside	—	—	—	—	2,678	3,228	3,759	5,034	6,547	8,611	10,042
Nova Scotia											
Amherst	—	—	3,781	4,964	8,973	9,998	7,450	8,620	9,870	10,788	10,551
Halifax	36,487	44,524	47,626	50,968	57,808	75,487	78,600	98,636	133,931	183,946	198,193
New Glasgow area[2]	—	5,998	9,184	13,488	19,303	24,668	23,571	24,444	27,157	26,942	27,310
Sydney	—	—	7,382	24,691	49,762	62,616	68,439	81,260	100,725	106,114	106,463
Truro	—	3,461	5,102	5,993	6,107	7,562	7,901	10,272	10,756	12,421	13,007
Newfoundland[3]							(1945)				
Corner Brook	—	—	—	—	—	—	—	13,243	18,223	25,185	27,116
Grand Falls–Windsor	—	—	—	—	—	—	—	—	—	—	14,113
St John's	—	—	—	—	—	—	—	59,474	67,749	90,838	101,161

1 Census of Canada. The figures refer to metropolitan areas and major urban areas for the larger centres and incorporated areas for the smaller ones.
2 New Glasgow area includes New Glasgow, Stellarton, Westville, Trenton, and Pictou.
3 Newfoundland became part of Canada in 1949. The figures for 1945 are derived from a Census of Newfoundland report.

British Columbia is strongly dominated by Vancouver, whose population is approaching one million, five times as large as that of Halifax. Hence, Vancouver contains nearly half of the provincial population, while Halifax has only one-twentieth of the Atlantic region population. Victoria, the second city of British Columbia, with 180,000 people, is slightly smaller than Halifax. Except for these two, British Columbia has no other cities exceeding 25,000 population. There are six between 15,000 and 25,000 in size, and four between 10,000 and 15,000. As the third largest city in Canada, Vancouver enjoys the self-sustaining growth typical of a large metropolis and is able to compete successfully with the cities of Central Canada. A hope for the Atlantic region is that it may develop its own metropolis, perhaps not on such a grandiose scale as Vancouver, but at least as a viable regional capital.

ORIGIN AND EVOLUTION OF THE URBAN SYSTEM

The urban system of the Atlantic Provinces is deeply rooted in the colonial past and has evolved through several centuries of change. The historic factor is especially important to the understanding of Atlantic Provinces cities because the region has been under European occupance longer than other parts of Canada, with the exception of the St Lawrence lowland of Quebec. The early settlement pattern was very complex owing to the prolonged conflict between the French and English in this battleground zone that became the Atlantic Provinces. They were locked in rivalry over the Newfoundland fishery, just as they were over the inland fur trade. But following Sir Humphrey Gilbert's claim for England in 1583, at least the St John's area was under English control thenceforth, except during a few French forays in the seventeenth century. Nova Scotia was first settled by the French at Port Royal in 1605, and Acadian settlement spread throughout the isthmus region and the Annapolis Valley. Although the English acquired sovereignty over the mainland of Nova Scotia in 1713, the first concerted effort to establish a strong settlement was the founding of Halifax in 1749. New Brunswick was also settled first by the French, both along the east coast and along the St John River valley. Most of the French settlers along the St John River subsequently were shifted to the east in order to make way for the English-speaking settlers. They arrived in large numbers following the American Revolution when the Loyalists sought refuge in this British territory. Prince Edward Island was settled by the French at an early date, but there also the French were largely replaced by settlers from England and Ireland after the British took possession in 1763 under the Treaty of Paris. The main nodes of settlement were established at places where good natural harbours existed, facilitating water transportation and giving access to fishing grounds.

St John's was the site of the earliest continuing settlement in Canada and perhaps in North America. A small, but remarkably well protected natural haven, St John's Harbour was frequented by many fishing vessels early in the sixteenth century. The earliest allusion to the name 'St John's' was in a letter written in 1527 by an English ship captain who mentioned that he had stopped at a harbour called 'St John's' and found numerous Norman, Breton, and Portuguese ships anchored there (Prowse 1895). St John's was situated near the Grand Banks fishing resource and was readily accessible to fishermen from western Europe. Hence, it became the seat of authority and a key fishing base. It grew slowly as the central place for Newfoundland, always retaining its primacy within the colony, and by 1867 it was reported as the third largest city in the Atlantic region, with a population of 22,000.

The deep and commodious Halifax Harbour was used as an anchorage by the French at an early date, but was never the site of a permanent settlement under their regime. At the prodding of the New England colonists the British government chose the site for development as a bastion of military power between New England and the French fortress at Louisbourg on Cape Breton Island. The town was established in 1749 with an initial settlement of 2500 people (Raddall 1965). Although it was essentially a military base, its size and importance from the beginning made it the government, trade, and commerce centre for the region. By the time of Confederation in 1867 Halifax had reached a population of about 36,000, making it the second largest city in the Atlantic region after Saint John.

The mouth of the St John River was recognized as a good harbour and strategic site by the French who built a fort there in 1631 (Raymond 1950). The fort was called St John because the river had been named the St John by Samuel de Champlain on 24 June 1604, the day of St John the Baptist. The settlement consisted of little more than a fort when the British captured it in 1758. British settlement was on a small scale until a large influx of Loyalists arrived from the United States in 1783. Like Halifax, Saint John was an 'instant town' with a population of several thousand people within a year or two of its founding. It thrived on trade and industry, particularly shipbuilding, and quickly became the largest city in the region, with a population of 52,000 by 1867. However, it lacked the colonial administrative function performed by Halifax and St John's because the capital of New Brunswick had been established at Fredericton, 60 miles inland, in a conscious effort to develop the interior.

Among the other sizable cities some were situated inland, but most owed their origins to their natural harbours. Settled by Loyalists, Sydney became the capital of Cape Breton when the island was designated as a separate province in 1784 and continued as such until the island was re-united with Nova Scotia in 1820. It enjoyed the advantage of a large harbour, through which coal was shipped, but it grew very slowly until the establishment of the iron and steel industry at the beginning of the twentieth century. The abundance of coal in the vicinity was a catalyst of this industrial development. Lying on the Petitcodiac River at the height of navigation for small vessels, Moncton was a very small community until the arrival of the railways, first a line from Saint John in 1860, then the Intercolonial Railway in the 1870s. Thenceforth, it became an important rail transportation centre. Charlottetown, situated on Hillsborough Bay, was laid out by the British in 1768 and became the Queens County seat and colonial capital (Clark 1959). The earlier French settlement, Port La Joie, near the mouth of the bay had been abandoned. Fredericton, origin-ally the site of an Acadian settlement, was surveyed and planned by the Royal Engineers for Loyalist settlers, and was named the New Brunswick capital. Other centres settled before Confederation, and in most cases before 1800, were Amherst, Truro, New Glasgow, Bathurst, Campbellton, and Edmundston. Only Grand Falls–Windsor and Corner Brook are of twentieth century origin and each is based upon the pulp and paper industry.

The urban system is characterized by remarkable stability, if not inertia. The main centres, once established, persisted in their dominant roles through a couple of hundred years, or more in some cases. Thus, the three largest cities in 1871, Saint John, Halifax, and St John's, remain the three largest cities a century later. They enjoyed the advantage of good natural harbours, an early start, and a thriving trade. St John's and Halifax also benefited from their governmental functions, while Saint John lacked this advantage. As transportation centres, Halifax and Saint John assumed national importance in the era of efficient land transportation after the railways were built.

The railways connected the Maritimes with Central Canada by a much firmer bond than existed previously and provided the basis for greater economic integration in the post-Confederation period. The Intercolonial Railway between Montreal and Halifax was completed in 1876 and the Canadian Pacific Railway line between Montreal and Saint John across the state of Maine was completed in 1890. These railway connections with Central Canada allowed Halifax and Saint John to develop as national ports, especially since they were ice-free in winter compared with the ice-blocked St Lawrence ports. The coming of the railways greatly aided Moncton as a central place and established it as the 'hub' of the Maritimes. The railway across Newfoundland built in the 1890s provided St John's with better access to the interior and the west coast region, but did not forge strong economic ties with Canada at this time.

Growth rates of urban centres in the Atlantic region were generally lower than those of cities in other Canadian regions during the last century. Halifax fared better than most, replacing Saint John by 1901 as the largest city in the region and quadrupling in size between 1901 and 1966. Reasonably well centred within its provincial territory, Halifax managed to acquire higher order functions and assure its primacy within Nova Scotia. The early dominance of Saint John was lost with the passing of its dominance in shipbuilding and trading in the era of sail. Its peripheral location within its provincial hinterland and its lack of significant government and military functions relegated Saint John to the slow growth category. St John's experienced a more rapid growth rate than its namesake and, despite its location on the easternmost periphery, retained its dominant central place function within the province. Growth of the Sydney area was fostered by the success of coal mining and the steel industry during the early part of this century. It has tapered off in recent years, owing to the economic difficulties of local industries. Moncton grew partly at the expense of Saint John by expanding its distribution functions in eastern New Brunswick and drawing Prince Edward Island within its tributary area. Fredericton and Charlottetown, on the other hand, were slow-growing provincial capitals with very limited tributary areas. Fredericton recently has spurted ahead of its island rival in population, chiefly as a result of expanding government and educational services in the larger province and the propinquity of the huge military base at Oromocto. Among the smaller places only a few industrial centres, including Corner Brook and Bathurst, have recorded substantial recent growth, along with Oromocto, which is a special case. Also, a large part of the population growth listed for Bathurst is accounted for by municipal boundary changes.

For the metropolitan areas information is available on the components of population growth during the period from 1951 to 1961 (Table 4.2). Although Halifax recorded the greatest increase in population, it fell short of the Canadian metropolitan area average by 4 percentage points. The overwhelming importance of natural increase in population growth is emphasized by the figures. Only in Halifax did net migration account for as much as one-third of the growth, while on the average in Canada half of metropolitan growth could be attributed to net migration.

Compared with that in other regions of Canada the process of urbaniza-

Table 4.2   Components of urban population growth from 1951 to 1961
(Source: Stone, Leroy O., *Urban Development in Canada*, p. 179)

Metropolitan area	Percentage change in population	Natural increase ratio	Net migration ratio	Relative importance of net migration
Halifax	37.3	24.6	12.7	34.1
Saint John	22.0	18.6	3.4	15.5
St John's	32.2	29.2	3.0	9.3
Average of 15 metropolitan areas in Canada	41.4	20.6	20.9	50.4

tion has proceeded more slowly in the Atlantic Provinces. According to the 1966 Census all of the Atlantic Provinces recorded a lower percentage of urban population than the Canadian average of 74 per cent. Nova Scotia, at 58 per cent, was the highest, followed by Newfoundland with 54 per cent, New Brunswick with 51 per cent, and Prince Edward Island with 37 per cent. During the period 1951 to 1966 a modest increase in urban population was accompanied by a substantial decline in the rural farm population (*Urban Centres in the Atlantic Provinces* 1969). For example, the urban population in New Brunswick was 43 per cent of the provincial total in 1951 and 51 per cent in 1966, while the comparable figures for the rural farm population were 28 and 8 per cent, respectively. A similar pattern prevailed in the other provinces. The rural non-farm population, already higher in the Atlantic Provinces than in other Canadian regions, registered a significant increase during this period. It is evident that the urban centres in the Atlantic Provinces are not attracting strong flows of rural population, as has been the case in other parts of Canada. One important factor is that large scale out-migration has taken place, Central Canada being the main destination. The Atlantic Development Board study of urban centres suggests two reasons why the rural non-farm population is overrepresented. First, many rural people have shifted from rural to urban occupations, but have remained in their rural dwellings and commute to their jobs. Secondly, many urban people have left the cities in search of relatively inexpensive housing in nearby rural areas. The urban population of the Atlantic Provinces is strongly concentrated in relatively small centres, and only one-quarter of the people live in cities of more than 100,000 population. In the whole of Canada about half of the population dwells in cities exceeding 100,000. The contrast in size between Vancouver and Halifax already alluded to is mute evidence of the disparity that exists.

URBAN PEOPLE AND CITY IMAGES

The cities of the Atlantic Provinces are distinctive in many respects. As a group they are set apart from the cities of other Canadian regions by their population characteristics and their physical appearance. At the same time, individuality exists within the regional context, as each of the major centres projects a unique character.

The slow growth of the cities and the sluggishness of the regional economy did not attract the immigrants who streamed into Canada in large numbers after World War II. The urban growth that has taken place can be attributed mainly to natural increase and migration from rural areas within the region. As a result, most of the people in Atlantic Provinces cities are Canadian-born. Among the seventeen metropolitan areas of Canada, Halifax, Saint John, and St John's ranked 14, 15, and 16, respectively, followed only by Quebec in percentage of population born outside Canada (Stone 1967).

With the exception of those in the Acadian areas of New Brunswick, most of the larger cities have populations predominantly British in origin. St John's, with 95 per cent British Isles origin, is by far the most British of all the metropolitan areas in Canada (Figure 4.2). Even Halifax, Saint John, and Sydney, in the 70–80 per cent range, exceed all other metropolitan areas except Victoria and London in this respect. Moncton, an important focus of Acadian culture, has a population that is more than one-third French in origin. Halifax has the most cosmopolitan population, nearly 17 per cent being of ethnic origins other than French or British. The Negro community in Halifax, numbering several thousand, traces its origins back to the war of 1812 when many refugees were brought from the United States (Winks 1971). To a certain extent ethnic origin is reflected in the affiliation to the various religious denominations. In general, the people of French and Irish descent belong to the Roman Catholic faith and those of English origins are Protestants. Hence, the St John's population, with a high proportion of Irish origin, is 50 per cent Roman Catholic, while the Saint John population, characterized by a dominance of English origin, is little more than one-third Roman Catholic. Saint John and Halifax are more than 60 per cent Protestant, while Sydney, St John's, and Moncton range from 40 to 50 per cent Protestant.

The basic features that, in combination, give substance to the distinc-

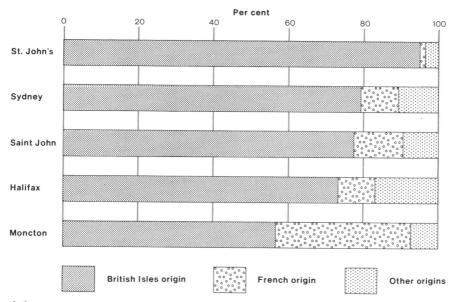

4.2

Distribution of Population among Three Ethnic Origin Categories 1961

tive image of Atlantic Provinces cities are the harbour, the sea, the fishery, the Old World character, the wooden buildings, the fortifications, and the lingering aura of history. Mostly seaports, these cities tend to focus on their harbours. The people have respect for the sea and look upon it as a valuable resource and a medium for the shipping trade. Their historic association with the fisheries gives them overtones of a rural, wholesome character. Even in the larger cities inshore fishermen still operate within the harbour or nearby, and large scale commercial fishing is based in some of them, St John's for example. Fishing in Saint John Harbour is conducted both from boats and with fish weirs, and a number of fishing lots are leased by auction every two years.

The Old World character is derived from the general appearance of central areas where the Victorian brick and stone business blocks are surrounded by row upon row of wooden dwellings with common walls and, in some cases, chimney pots. In these residential sections the narrowness of the streets is emphasized by the crowding of buildings that rise steeply at the sidewalk line. The old houses, with their high ceilings and three storeys, are dominated by vertical lines. Their wooden construction contrasts with the brick and stone so common in Central Canada. On the whole, the buildings are not particularly old, being late nineteenth century or early twentieth century for the most part. This is so because extensive conflagrations periodically destroyed central areas during the nineteenth century and the Halifax explosion of 1917 resulted in unbelievable devastation. Visual evidence of past occurrence of fires can be observed in some cities. In Saint John, for example, there are many business blocks and bank buildings carrying dates on their facades of 1878 or 1879, suggesting that widespread building replacement occurred following a fire, which was the case. Disastrous fires that destroyed a large part of the city occurred in St John's in 1846 and in 1892 (Pearson 1969). After the 1892 fire the business blocks were rebuilt in brick, but wood again was used for the row houses. The slow growth of the cities and the accompanying slow replacement of buildings in central areas are characteristics that help to account for the elderly appearance of these areas. The building of tall, new office blocks in downtown areas is beginning to change the skylines and streetscapes, but this process has not advanced so far as in other Canadian cities. In addition, the slower pace of life in the Atlantic Provinces cities does not contradict the Old World character. The numerous fortifications on prominent sites, along with the old buildings, create cityscapes virtually in the form of historic settings. The visual impact of such cities conveys a sense of history that leavens the effect of modernity.

Though displaying the regional characteristics, the individual city proclaims its uniqueness in many ways. Several examples among the larger centres will serve to illustrate this point. Thus, Halifax not only is focused on its harbour and associated port activities, but also revolves around its huge naval base. First and foremost it is a navy town. Its roots in the past are prominent features; the Citadel dominates the core area from its commanding site above the harbour, while fortified hills, points, and islands guard the harbour entrance. These stand as symbols of the historic military and strategic role of Halifax. Against the background of its historic character Halifax displays elements of modernization in the form of circulation patterns for vehicular traffic and numerous high-rise apartment buildings. Scotia Square also provides an ultramodern focus for the urban core. As the largest centre east of Quebec, it has acquired some of the rudiments of a big city.

In keeping with the regional identity, Saint John exudes an Old World character, especially in the core area where there has been less renewal than has occurred in Halifax. The massive stone and brick buildings of Victorian architecture dating from the 1870s portray the Saint John of a century ago when it was at the height of its trading and financial success.

The extensive Loyalist Burial Ground in the centre of the city, bordered by King's Square, laid out in the form of a Union Jack, is a stark reminder of the city's past and the loyalty of its founders. But Saint John is an industrial city also, with its pulp and paper mills, its oil refinery, and its shipyard ringing the central area. It lacks the naval function, but its commercial docks and railway yards dominate the waterfront of the small inner part of the harbour. It has the aspect of an old city that earns its keep in the modern world through honest labour.

In St John's the land-locked harbour is truly the focal point of the city, acting like a hub from which the streets fan out like the spokes of a wheel. The steep slopes, funnelling down to the harbour, on which the older part of the city is built heighten the effect of the harbour's influence. Although the harbour is the scene of commerce, its association with the fishing industry is emphasized by the presence of fishing vessels. It is used as a provisioning base by vessels of many countries operating on the Grand Banks, and during storm warnings as a safe haven (Forward 1967). The Battery area near the mouth of the harbour has the appearance of an outport fishing village, the small wharves and sheds festooned with nets. These trappings of the fishery link the modern city in the eyes of the beholder inescapably with its historic past. St John's also has its old buildings and row houses that complete the Old World, fishing port image. The newer, suburban areas of St John's, as is the case in other Atlantic cities, are imposed as a veneer of 'Canadian modern' upon the historic community.

URBAN FUNCTIONS

Delving beneath the observable images, it is worthwhile attempting to measure the significance of the different urban functions. A common approach to the study of urban functions has been through the determination of the urban economic base. Detailed economic base studies of individual cities have been carried out by planning or research agencies, such as the Fredericton–Oromocto study by the Atlantic Provinces Economic Council (*A Service-based Community: The Example of Fredericton–Oromocto* 1969) and Sinclair's Halifax study (1961). Certain functional aspects of cities have been studied, for example trade centres and distribution functions by the Atlantic Development Board (*Urban Centres in the Atlantic Provinces* 1969), and financial functions by Kerr (1965). A comprehensive functional classification of Canadian cities was carried out by Maxwell (1965), who applied the minimum requirements method of determining the urban economic base suggested by Ullman and Dacey. Maxwell's approach is the most useful for the purpose of achieving a general measure of urban function on a comparative basis. A limitation of his study is that it was based on employment statistics published in the 1951 Census. The most recent data available are in the 1961 Census because the required statistics were not collected in 1966. In order to update one aspect of Maxwell's findings, the minimum requirement in each industry group was calculated within size categories of Canadian cities based on 1961 figures.

In general, the urban centres of the Atlantic region are classified by Maxwell as regional capitals, with manufacturing relatively unimportant. In other words, central place functions assume greatest importance in their economies. This is typical of the urban centres in the peripheral regions of Canada, as opposed to those of southern Ontario and Quebec, the Canadian 'heartland.' The larger cities have diversified economies, with a few specialized exceptions such as Halifax and Sydney. Many of the smaller cities, on the other hand, are dominated by industrial or mining activities based on local natural resources.

Tables 4.3, 4.4, and 4.5 present relevant statistics based on labour force by industry group as reported in the 1961 Census of Canada. All Atlantic Provinces cities that were over 10,000 population at that time are included. The minimum percentage of employment in each industry group was determined within several size categories of Canadian cities. These minima are listed under the heading 'minimum requirement' and are equated with the service component of the urban economy. In the case of a particular city, the percentage of employment in each industry group in excess of the minimum is equated with the basic component of the urban economy. Presumably, this basic employment is engaged in activities which bring income from outside the community, while the service employment is supported by funds generated within the community. Accordingly, these data provide a rough guide, in terms of labour force, of the economic importance of the various industry groups.

Among the four largest cities Halifax and Sydney are highly specialized, while Saint John and St John's are much more diversified (Table 4.3). Nearly one-third of the labour force in Halifax is engaged in public administration and national defence functions. The naval base accounts for about two-thirds of the Canadian naval forces and the naval dockyard employs 1700 civilians, along with the uniformed personnel. Sydney's specialization in the primary industry of coal mining is indicated by the high percentage of labour force recorded in 1961. Since that time, however, many mines have been closed and the economy is being oriented more toward manufacturing and service activities. Manufacturing plays a greater role in the economies of Saint John and Sydney than in those of Halifax

Table 4.3 Percentage of employment, minimum requirement, and excess employment for Atlantic Provinces cities over 100,000 population, 1961

Industry group	Minimum requirement derived from the 14 Canadian cities of population 90,000–194,000	Halifax		Saint John		St John's		Sydney	
		Employment	Excess employment	Employment	Excess employment	Employment	Excess employment	Employment	Excess employment
Agriculture, forestry, fishing and mining	0.5	0.5	—	1.1	0.6	2.0	1.5	24.2	23.7
Manufacturing	9.4	10.4	1.0	20.6	11.2	9.4	—	16.6	7.2
Construction	3.8	4.7	0.9	6.8	3.0	7.0	3.2	3.8	—
Transportation, communications and utilities	4.9	10.7	5.8	14.4	9.5	14.3	9.4	10.7	5.8
Wholesale trade	2.9	5.7	2.8	8.8	5.9	7.7	4.8	3.5	0.6
Retail trade	11.6	11.6	—	13.2	1.6	16.8	5.2	12.5	0.9
Finance, insurance and real estate	2.0	4.2	2.2	4.2	2.2	3.1	1.1	2.0	—
Community, business and personal services	16.4	21.6	5.2	23.7	7.3	24.5	8.1	19.7	3.3
Public administration and defence services	3.2	30.6	27.4	7.2	4.0	15.2	12.0	7.0	3.8
Total	54.7	100.0	45.3	100.0	45.3	100.0	45.3	100.0	45.3

Table 4.4 Percentage of employment, minimum requirement and excess employment for Atlantic Provinces cities between 15,000 and 60,000 population, 1961

Industry group	Minimum requirement derived from the 25 Canadian cities of population 27,000-84,000	Moncton		Minimum requirement derived from the 24 Canadian cities of population 15,000-26,000	Corner Brook		Fredericton		Charlottetown	
		Employment	Excess employment		Employment	Excess employment	Employment	Excess employment	Employment	Excess employment
Agriculture, forestry, fishing and mining	0.4	0 8	0.4	0.4	2.1	1.7	2.3	1.9	0.9	0.5
Manufacturing	4.8	11 7	6.9	6.2	30.1	23.9	8.4	2.2	10.2	4.0
Construction	4.8	5 4	0.6	4.5	7.9	3.4	5.8	1.3	7.8	3.3
Transportation, communications and utilities	4.3	20 0	15.7	5.2	13.5	8.3	9.8	4.6	12.5	7.3
Wholesale trade	1.8	8 7	6.9	2.4	5.6	3.2	5.8	3.4	6.6	4.2
Retail trade	10.0	18 3	8.3	11.2	16.8	5.6	14.1	2.9	15.6	4.4
Finance, insurance and real estate	2.1	4.1	2.0	2.0	2.0	—	3.3	1.3	3.9	1.9
Community, business and personal services	15.3	21.6	6.3	16.9	17.8	0.9	28.1	11.2	32.3	15.4
Public administration and defence services	2.7	9.4	6.7	3.4	4.2	0.8	22.4	19.0	10.2	6.8
Total	46.2	100.0	53.8	52.2	100.0	47.8	100.0	47.8	100.0	47.8

Table 4.5 Percentage of employment, minimum requirement and excess employment for Atlantic Provinces cities between 10,000 and 15,000 population, 1961

Industry group	Minimum requirement derived from the 33 Canadian cities of population 10,000–15,000*	Edmundston		Truro		Amherst		Oromocto	
		Employment	Excess employment	Employment	Excess employment	Employment	Excess employment	Employment	Excess employment
Agriculture, forestry, fishing and mining	0.7	2.9	2.2	1.7	1.0	2.0	1.3	0.2	—
Manufacturing	2.4	27.3	24.9	18.2	15.8	25.9	23.5	0.4	—
Construction	3.0	4.4	1.4	5.4	2.4	7.8	4.8	0.3	—
Transportation, communications and utilities	3.3	13.1	9.8	14.7	11.4	8.1	4.8	0.5	—
Wholesale trade	0.8	2.3	1.5	7.8	7.0	4.9	4.1	0.2	—
Retail trade	9.8	13.4	3.6	15.9	6.1	16.9	7.1	2.4	—
Finance, insurance and real estate	1.5	2.7	1.2	4.2	2.7	4.1	2.6	0.8	—
Community, business and personal services	12.7	28.0	15.3	24.3	11.6	24.6	11.9	3.7	—
Public administration and defence services	2.7	5.9	3.2	7.8	5.1	5.7	3.0	91.5	—
Total	36.9	100.0	63.1	100.0	63.1	100.0	63.1	100.0	—

*Oromocto was omitted from the group used to determine minima because it was not considered a viable, independent city.

and St John's. Indeed, St John's records the lowest percentage in manufacturing of all Canadian cities in its size group and Halifax registers a figure only slightly higher. The port and railway terminus functions account for the prominence of the transportation, communications, and utilities category in the urban economies of all four cities. Wholesale trade, which is indicative of the importance of a city as a central place, is well developed in all except Sydney. The Cape Breton hinterland of Sydney is rather sparsely populated. As financial centres Halifax and Saint John appear to be on a par, but Halifax has more head offices of insurance, loan and trust companies than has Saint John. Although it lacks a stock exchange, Halifax is the only Atlantic Provinces city to be ranked among the top fifteen cities performing financial functions in Canada (Kerr 1965). Community, business and personal services, including educational, medical and religious, assume considerable importance in the three leading central places, while playing a lesser role in Sydney. The universities and major hospitals are salient features of this service component.

Moncton stands alone as the only Atlantic Provinces representative among the twenty-five Canadian cities in the population range of 27,000 to 84,000. Its characterization as a prominent central place and transportation centre is borne out by the employment figures (Table 4.4). In fact, Moncton registers the highest percentage of labour force in both transportation and retail trade among the twenty-five cities.

In the next group Fredericton and Charlottetown reveal similar economy profiles in their emphases on public administration and services (Table 4.4). Their provincial capital functions, provincial universities, and services for local tributary areas occupy substantial portions of their labour forces. The great dominance of government functions in the relatively small city of Fredericton is indicated by the fact that the percentage in public administration in Fredericton exceeds that of any other city in the size group. The economic importance of the provincial government in Fredericton is paralleled by the recent expansion of federal government job opportunities in that area. The nearby Canadian Forces Base Gagetown has benefited Fredericton, as well as Oromocto. Charlottetown leads the group in percentage of employment in wholesale trade. Corner Brook, on the other hand, is important for manufacturing, as well as for its central place functions. The Bowaters Newfoundland pulp and paper mill is the leading industrial employer. Transportation is a significant function in all three urban economies.

In the final group of thirty-four cities (Table 4.5) Oromocto is unique in being a recently established dormitory community on the edge of Canada's largest army training base. With over 90 per cent of its labour force in the public administration and defence services category, Oromocto cannot be classified as a viable, independent urban centre comparable with other Canadian cities of similar population. Hence, it was omitted from the group in which minimum requirements were calculated. Among the others Truro has a well-balanced economy in which central place functions occupy a prominent position, while Edmundston and Amherst are more specialized in manufacturing activities, with over one-quarter of total employment in that category.

The following places could not be included in Table 4.5 because the relevant statistics were not available in the 1961 Census; Bathurst, Campbellton, Summerside, and Grand Falls–Windsor. Labour force by industry group was published only for cities of 10,000 population or over. Bathurst and Grand Falls–Windsor actually exceeded this size, but were not incorporated as such until after the 1961 Census. Both of these centres have large pulp and paper industries as bases for their economies. Campbellton and Summerside are more diversified, but the latter has considerable economic dependence upon the Canadian Forces air base situated there.

Central Places and Urban Interaction

A more thorough consideration of central place functions serves to illuminate the nature of the urban hierarchy and the intraregional fragmentation that exists. An investigation of trade centres and trade areas was carried out recently by the Atlantic Development Board (*Urban Centres in the Atlantic Provinces* 1969). The cities were classified into types of centres on the basis of numbers of wholesale and retail business functions performed (Table 4.6). Forty-four functions were selected for the purpose of classifying centres.

The primary wholesale-retail centres are the five largest cities and they perform all of the forty-four selected functions. Although metropolitan area population figures were used for the first three, the 'major urban area' population totals were not used for Moncton and Sydney. Accordingly, these two cities actually constitute higher proportions of their trade area populations than indicated. Even with the use of the major urban area figure, it is apparent that both St John's and Moncton stand apart from the others in the enjoyment of trade areas that contain nearly twice as many people as are resident in the cities themselves. This emphasizes the importance of central place functions in St John's and Moncton.

Among the secondary wholesale-retail centres Truro is distinguished by its remarkable range of functions in relation to its small size. Apparently, it is able to compete successfully in its limited trade area with the higher-order centre of Halifax situated only 60 miles away. New

Table 4.6 Trade centres and trade area population, 1961
(Source: *Urban Centres in the Atlantic Provinces*
Atlantic Development Board, Ottawa, 1969, p. 47)

Type	Number of selected business functions (total 44)	Centre	Population	Estimated population of trade area	Percentage of trade area population in trade centre
Primary wholesale–retail	44	Halifax	183,946	244,345	75.2
		Saint John	95,563	148,090	64.5
		St John's	90,838	244,825	37.1
		Moncton	43,840	147,150	29.8
		Sydney[1]	77,052	145,416	53.0
Secondary wholesale–retail	28–31	Charlottetown[2]	21,663	87,221	24.8
		Corner Brook	25,185	87,862	28.6
		Fredericton	19,683	80,417	24.5
		New Glasgow[3]	22,408	43,908	51.0
		Truro	12,421	32,909	37.7
Complete shopping	21–27	Amherst	10,788	37,767	28.5
		Bathurst[4]	5,494	39,856	13.9
		Campbellton	9,873	23,775	41.5
		Edmundston	12,791	43,371	29.5
		Grand Falls–Windsor	12,111	65,281	18.5
		Summerside	9,613	44,410	21.6
Partial shopping	16–20	Oromocto	12,170	14,112	86.2

1 Sydney includes North Sydney, Glace Bay, and New Waterford.
2 Charlottetown includes Parkdale and Sherwood.
3 New Glasgow includes Stellarton, Westville, and Trenton.
4 Bathurst municipal boundary was greatly enlarged subsequent to 1961 and included over 15,000 people in 1966.

Glasgow has a rather small trade area population in relation to its size, especially when Pictou is considered as part of the dispersed city population. The other three centres serve extensive regional tributary areas. In the complete shopping group Bathurst is an anomaly because its true population would constitute considerably more than 14 per cent of the trade area total. Finally, Oromocto falls in the partial shopping category because its commercial functions are poorly developed relative to its size. It is likely that many of the goods consumed by the community are purchased from the retail outlets in the military base.

Another undertaking of the Atlantic Development Board study was an examination of the distribution function in terms of transportation costs. Based on the premise that internal distribution costs are to be minimized, optimum allocations of distribution centres were derived for successive cases of one centre, two centres, three, and so on to a total of eleven. The size of the market segment used in the calculations was assumed to be proportional to population. With only one centre for the Atlantic Provinces Truro is the optimum choice, but with two centres it is replaced by Moncton and St John's. Obviously, the Maritimes and Newfoundland, basically, are two separate trading areas, the former being three times the size of the latter. When three centres are permitted Halifax joins the group and three provincial markets result, Prince Edward Island being part of the Moncton area. With four centres the Newfoundland market is split between St John's and Corner Brook, by far the larger area being tributary to Corner Brook. The introduction of five centres splits the Nova Scotia market into the Cape Breton Island portion served by Sydney and the mainland served by Halifax. Only when more than five centres are allowed does the New Brunswick market split into two, with Fredericton added as a sixth centre, and into three when Bathurst is added as a seventh. It was discovered that between 90 and 95 per cent of the ultimate distribution cost reduction is achieved by using no more than five to seven centres.

The proportions of the total Atlantic Provinces population allocated to centres in the various cases postulated are indicated in Table 4.7. It is noteworthy that Saint John, in spite of its size, does not appear in the list until nine centres are used. In reality, Saint John has overcome serious disadvantages of distance to achieve its prominent role as a central place. Under present conditions it is apparent that Halifax is the closest approach to a primate trading centre in the Atlantic Provinces region. It is by far the largest city and with six or more distribution centres, as is the case, it has the largest proportion of the regional population within its distribution area. On the other hand, when fewer than six centres are postulated Moncton captures the largest percentage.

Table 4.7   Percentage of total population allocated to centres
(Source: *Urban Centres in the Atlantic Provinces*
Atlantic Development Board, Ottawa, 1969, p. 62)

Centre	Number of centres postulated						
	1	2	3	4	5	6	7
Truro	100						
Moncton		75	38	38	39	24	16
St John's		25	25	12	12	12	12
Halifax			37	36	26	26	23
Corner Brook				14	14	14	14
Sydney					9	9	9
Fredericton						15	15
Bathurst							11
Total	100	100	100	100	100	100	100

An indicator of economic, governmental, and social interaction is the volume of air passenger traffic between cities. The following observations are based upon the origin and destination statistics for 1969 (*Air Passenger Origin and Destination* 1969). These figures indicate the total number of air passenger trips between cities, including both directions of travel, but omitting reference to intermediate ports of call (Figure 4.3).

A large proportion of the total air passenger volume is accounted for by the five largest cities. Halifax emerges as the key air transport centre of the region, having by far the largest volume of traffic among the major cities. Also, its interaction with Montreal and Toronto reaches a volume that is almost equal to that of the other four cities combined. That Halifax is not a primate city of the region, however, is borne out by the substantial volumes of traffic between the other four cities and the inland metropolises. St John's and Saint John have passenger interchanges with each of Montreal and Toronto amounting to totals very similar to those of their interchanges with Halifax. In the case of Moncton, the interchange with Montreal and Toronto greatly exceeds that with Halifax. However, it is so readily accessible to Halifax by road and rail that volume of air traffic between these cities may greatly under-represent their true interaction. Similarly, air traffic between Moncton and Saint John is insignificant, owing to the ease of land transport connections. Only Sydney appears to interact more strongly with Halifax than with the Central Canada cities, an indication of the strength of central place interrelationships within provincial tributary areas. It is interesting to note that Toronto slightly exceeds Montreal in volume of air traffic with these Atlantic Provinces

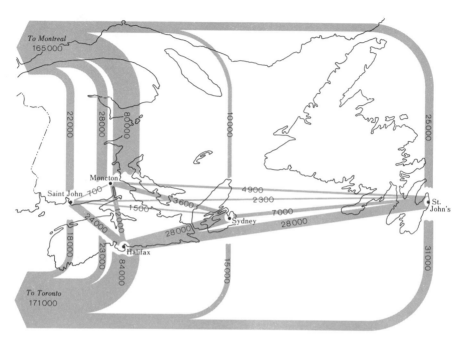

4.3

Number of Air Passengers Travelling between Selected Cities 1969

cities, although it is the smaller of the two national metropolises and is the
more distant. Also, the New Brunswick cities that are closest to Montreal
appear to have a stronger interaction with that centre, while the others
that are more remote exhibit a bias towards Toronto.

The statistics relating to smaller centres tend to support the generalizations already made (Table 4.8). Data are available for Fredericton,
Charlottetown, and Stephenville, the airport for Corner Brook. The
orientation towards Toronto, Montreal, and Halifax is well developed,
being in fact a parallel orientation in the cases of Charlottetown and
Stephenville. Fredericton displays the strong interaction with the inland
centres at the expense of Halifax that is characterized by Moncton, but to
a lesser degree. Among the smaller Atlantic Provinces cities, as was the
case with the larger ones, interaction with Montreal declines eastward,
while the reverse occurs with Toronto. The strength of the intraprovincial
tie is illustrated by the high volume of traffic between Stephenville and St
John's. Fredericton, of course, interacts with its higher-order provincial
centres of Saint John and Moncton mainly through land transport modes.

Table 4.8   Number of air passengers travelling between selected small cities and large cities, 1969*
(Source: *Air Passenger Origin and Destination* 1969, Domestic Report, Dominion Bureau of Statistics, Ottawa, 1970)

	Toronto	Montreal	Halifax	Saint John	Moncton	Sydney	St John's
Fredericton	17,000	22,000	14,000	—	1,000	1,000	2,000
Charlottetown	5,000	5,000	5,000	—	2,000	—	1,000
Stephenville	5,000	4,000	5,000	—	1,000	3,000	16,000

*The figures are rounded to the nearest thousand and numbers below 500 are omitted.

INTERNAL STRUCTURE OF CITIES

Inside the city the varied land use, socio-economic, and political patterns are responses to functional demands and have evolved through historical adjustments to site and situation. Although every city is unique, there are many structural elements that are similar from one city to another because similar functions beget similar responses. Most of the cities are characterized by a sea orientation with a natural harbour as a focal point. Those inland are generally smaller and situated on rivers. From a point of origin, usually on the harbour shore, a street system was established and settlement spread along the harbour and inland. In many cases there were several points of origin, resulting in the development of complex morphologies. This occurred particularly in Halifax, Saint John, and Sydney, where extensive harbour embayments offered numerous different sites adapted to different functional needs. In Sydney there was the additional factor of scattered coal mine locations to induce the development of a multiple-nuclei community. Those cities with small harbours, such as St John's, or situated inland, such as Fredericton, tend to be far more compact in structure. Of course, in common with other North American cities, considerable suburbanization has occurred in all of the Atlantic Provinces cities, especially since World War II.

LAND USE PATTERNS

The use of urban land can be generalized into a few major categories to facilitate comparison: transportational, industrial, commercial, institutional, residential, and recreational.

Transportational
The harbour focus of transport systems is a noteworthy feature of the urban circulation patterns. Central business districts are situated near the harbour shores amid grid patterns of streets conforming with shoreline alignment, the main business street being parallel with the shore in most cases. These downtown areas, in conjunction with the port facilities developed nearby, have acted as focal points for transportation. On the minus side, the harbours exist as land transportation barriers in some cities, requiring ferries and bridges, as in Halifax and Saint John. The major bridges are integral parts of modern, limited access routes, though most of the circulation patterns in Atlantic Provinces cities are composed of conventional streets and thoroughfares with frequent intersections. Few

of the cities are large enough to require expressways and limited public works budgets generally have not permitted expressway development.

Railways in the port cities naturally focus on the port facilities. Extensive railway yards with many miles of trackage border the deep-sea shipping terminals on Halifax and Saint John harbours, while more limited railway terminal facilities exist in the smaller ports. Moncton has an elaborate development of railway yards in association with the Canadian National maintenance shops and the rail car classification function performed there. Railway lines frequently have pre-empted harbour shorelines in wending their way to the shipping terminals, their builders having sought easy grades for rights of way. This has created an unfortunate barrier to public access and to possible use of waterfront zones adjacent to residential areas as parks. Halifax and Saint John are hampered most by this effect. From an industrial point of view, however, the harbour-bordering rail network is advantageous.

Port facilities are most extensive in Halifax and Saint John, the only cities performing national port functions. The other ports are chiefly regional in function, handling general and bulk cargo for their provincial or sub-provincial hinterlands (Forward 1967, 1969). Extensive facilities in Halifax and Saint John include numerous conventional general cargo wharves and transit sheds, as well as specialized container and bulk handling terminals. These ports have long been known as 'winter ports' because they were engaged in the general cargo trade on a large scale only in winter when the St Lawrence ports were closed by ice. During the past decade, however, icebreaking activities and the wider use of ice-strengthened ships have permitted year-round operations at Quebec and Montreal. As a result, Halifax and Saint John recently have faced increasingly stronger competition in the conventionally handled general cargo trade. The building of new container terminals at the beginning of the 1970s in Halifax and Saint John, and the inauguration of full container services to these ports has come in time to counteract the diversion of other types of cargo. The year-round regularity of container business will be a great boon to the Atlantic ports, enabling them to live down their reputations as merely 'winter ports.'

Industrial
The port zones, generally, are characterized by industrial development, many industries occupying waterfront sites and having their own wharves. Most of the principal industries are in the primary rather than secondary category and often are based on raw materials assembled by water trans-

portation. For example, on waterfront sites Halifax has two petroleum refineries, Saint John has one petroleum refinery, a sugar refinery, and two pulp and paper mills, and Sydney has an integrated iron and steel industry. With a few exceptions, secondary industries are mostly situated inland. The notable exceptions on waterfront sites are the large shipbuilding and ship repair industries in Saint John and Halifax, and the Volvo automobile assembly plant in Halifax. At inland locations small areas of predominantly secondary industries exist in various cities, though only Saint John and Sydney have extensive industrial development. Many cities have zoned large areas for industry and great efforts are being made to attract new manufacturing activities. Most of the larger cities have established industrial parks, usually municipally owned and operated. The provision of services in these parks and the institution of site planning, design coordination, and landscaping produces more attractive industrial zones than in the haphazardly developed older areas.

Commercial

The typical North American pattern of commercial structure has prevailed in the Atlantic Provinces cities, the historic central business district and outlying districts or business streets being joined more recently by integrated shopping centres of various sizes and types. These range from the neighbourhood centres with supermarkets as cores to regional centres focusing on department stores. Most of the larger cities boast at least one fully enclosed shopping centre, such as Halifax Shopping Centre, Highfield Square in Moncton, and Avalon Mall in St John's. Wholesale establishments that were formerly situated mainly in old buildings around the edges of the central business districts have tended to relocate in outlying areas with good transportation facilities, often in close association with the service manufacturing industries.

A pronounced decline of the central business district occurred during the shopping centre boom in Halifax and Saint John, and to a lesser extent in some of the other cities. In Halifax a Simpsons department store was established in a strategic location on the isthmus of the peninsula many years ago (Figure 4.4). The accessibility of the isthmus site from the Dartmouth side of the harbour was greatly enhanced with the building of the Angus L. MacDonald Bridge in the 1950s. The central business district was further weakened in the early 1960s when the only remaining major department store, Eaton's of Canada, shifted to the Halifax Shopping Centre on the isthmus adjacent to Simpsons. By the mid-1960s the historic central business district was relegated to the status of a regional shopping centre, while the isthmus zone functioned as the retail core of

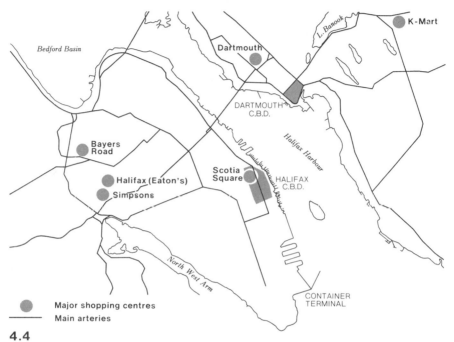

4.4

Major Shopping Areas in Halifax

metropolitan Halifax. Another factor in the decline of the central business district was the lack of adequate parking facilities. No municipal parking garages were built and there were few such private facilities. The situation had changed radically by the end of the decade. The Scotia Square shopping plaza with its associated parking garage was opened, containing two department stores and a large number of specialty stores. Accordingly, the central business district at last is in a position to compete with the shopping centres, despite the recent establishment of a second harbour crossing even farther away from downtown than the original bridge.

Renewal of retail facilities in the central business district of Saint John was minimal during the rise of the shopping centres and there was little expansion of parking capacity. Although the major department store did not depart, as happened in Halifax, downtown retailing tended to mark time, rather than forge ahead towards the creation of a more dynamic urban core. In St John's the central business district countered the shopping centre challenge with the temporary establishment, on an experimental basis, of a three-block-long, traffic-free, and landscaped mall on Water Street. Parking problems were not entirely solved, but the amenities for shoppers were greatly improved. Moncton's Highfield Square was built at one end of the main downtown shopping area, thereby strengthening rather than weakening the central business district.

Institutional
Military, governmental, educational, medical, and religious uses of land are placed in this category. As major central places, the larger cities have considerable space devoted to institutional uses. The patterns of use differ considerably from one city to another, but the older, central areas generally contain the greatest range of significant institutional functions. Military land use in Halifax is extensive both in the central area, where the naval headquarters, dockyard, and training establishment are situated, and at various points bordering the harbour on the Dartmouth side, where the naval battery, air station, and other installations are found. The universities also occupy large acreages in suburban areas in Halifax, Saint John, St John's, and Moncton. In the case of the Memorial University of Newfoundland the new campus on Confederation Parkway is the result of a relocation from a smaller site closer to the city centre.

Residential
Suburbanization in the postwar period has greatly altered the pattern of residential use that existed previously. The newer areas of predominantly single-family, detached dwellings contrast with the older areas of wooden row houses and three-storey flats. All of the suburban areas are not of high quality, however. Many areas of rocky, hilly terrain are characterized by haphazard development and incomplete services. In the extensive older areas of Atlantic Provinces cities housing quality has declined seriously, many of these dwellings being in dire need of renovation or complete renewal. At the other end of the housing spectrum, modern high-rise apartments have been introduced to some of the older housing areas, particularly in the largest cities. Halifax has the greatest number of tall apartment blocks, but generally speaking, high-rise apartment development has advanced on a lesser scale than in Central Canada.

There is considerable variation from city to city in the proportions of dwelling units falling into each of the three basic structural types (Table 4.9). In general, the larger cities, with the exception of Sydney, tend to have more apartments and single attached dwellings and fewer single detached than the smaller cities. The many dispersed components of Sydney tend to resemble the smaller cities of the region in this respect, giving an aggregate figure for the single detached category that is especially high. Home ownership percentages, in most cases, are closely in accord with

Table 4.9 Structural types and ownership status in percentages of total dwelling units, 1966*
(Source: *Census of Canada, 1966*)

City	Single detached	Single attached	Apartments and flats	Owned	Rented
Halifax	52	7	41	52	48
Sydney	73	13	14	79	21
Saint John	41	7	52	47	53
St John's	51	22	27	68	32
Moncton	60	8	32	60	40
New Glasgow area	75	10	15	70	30
Corner Brook	75	7	18	77	23
Fredericton	54	5	41	53	47
Charlottetown	44	24	32	46	54
Bathurst	70	7	23	64	36
Grand Falls–Windsor	81	7	12	84	16
Oromocto	37	55	8	8	92
Truro	60	6	34	61	39
Edmundston	55	9	36	59	41
Amherst	64	13	23	64	36
Campbellton	62	9	29	61	39
Summerside	57	16	27	54	46

*The figures for Halifax, Saint John, and St John's refer to metropolitan areas, for Sydney and Moncton to major urban areas, and the remainder to cities, the New Glasgow area being the five centres consolidated.

proportions of single detached dwellings. Regionally, home ownership is very high in the Newfoundland cities and low in the Prince Edward Island cities. In Nova Scotia only Halifax registers an ownership figure of less than 60 per cent, chiefly owing to its high proportion of apartments and flats. In New Brunswick both Saint John and Fredericton also have many apartments and Oromocto, as usual, is a special case. In this dormitory community over 90 per cent of all dwellings are rented and more than half of them are double or row houses. Among all the Atlantic Provinces cities over ten thousand population, Oromocto has the lowest proportion of single detached houses.

Recreational

Public open space and park land is more abundant in some cities than in others, but on the average Atlantic Provinces cities are comparable with those elsewhere in Canada in availability of such space. Owing to the relatively small size of most of the cities, the inhabitants enjoy the advantage of having easy access to the open country which offers varied recreational opportunities. The rather limited extent of waterfront parks in the coastal cities is an unfortunate shortcoming in this maritime environment. Also, many of the old residential areas are poorly provided with playgrounds and other types of parks. On the positive side, some fairly large park areas have been set aside in the newer parts of the cities. Halifax, in particular, has benefited by the establishment of parks at the site of historic fortifications that are no longer required for military purposes.

SOCIO-ECONOMIC PATTERNS

Variations in housing quality and value within a city and variations in educational and income levels within the population are important indicators of general socio-economic patterns. Using the most recent available census tract statistics for Halifax, Saint John, and St John's, it is possible to compare the socio-economic characteristics of these cities on a uniform basis (Figure 4.5). Unfortunately, such data are unavailable for other Atlantic Provinces cities, and are incomplete for Saint John.

The three criteria on which the socio-economic districts have been defined can be stated explicitly. The first indicator is the percentage of the population not attending school when the census was taken that had one or more years of university training. The second is the percentage above or below the city's median value of owner-occupied dwellings, and the third is the percentage above or below the city's average wage and salary income per family. The categories mapped were arbitrarily chosen to high-

The Saint John map is based only on university training and family income because data for dwelling value are incomplete in the census tract bulletin

4.5

Socio-Economic Districts by Census Tracts in Halifax, Saint John and St. John's 1961

light the extremes among the socio-economic districts. It is regrettable that the large size of many census tracts tends to mask the finer details of the real distribution.

The lowest level socio-economic districts appear to be in the older parts of the cities, although the north end of Dartmouth is an exception. Saint John is unusual in having four separate districts at the low end of the scale. The tracts bordering North West Arm stand out as the highest zone in metropolitan Halifax, while the highest level areas in St John's and Saint John are inland. In Saint John other high level districts exist within

the huge northern tract, but they are separated by areas of lower status.
St John's has the simplest pattern, with the lowest category surrounding
the harbour, the highest in a northern block and the average on the west
side. Most of the urban renewal projects undertaken in these cities have
been in the areas marked as low socio-economic level.

Housing quality and community facilities vary considerably from one
city to another within the Atlantic Provinces, reflecting such factors as age
of the community, nature of the economic base, and level of prosperity.
The Atlantic Development Board study (1969) presented both housing
and community facilities indexes based on unpublished 1961 census
material from the Dominion Bureau of Statistics (Table 4.10). The variables used to determine the housing index were as follows: houses in need
of major repair, in need of minor repair, with furnace heat, with toilet,
houses connected with municipal mains, with sewer connections, and
houses with hot and cold water. Low values in the index indicate high
quality of housing, and vice versa. The community facilities index could
not be based upon census information, but other information was used to
derive the following variables: percentage of streets paved, frequency of
garbage collection, National Board insurance rating, pupils per teacher,
and 1963–5 neonatal deaths and stillbirths. Rearrangement of the cities
that fall within the metropolitan area groupings used in the present study
has been carried out, but the index values have been left unchanged. With
a few exceptions, ranking of cities is similar in both housing quality and
community facilities lists. Oromocto leads both lists because it is an
entirely new and fully planned city that has not had time to deteriorate.
Most of the others in the top group are prosperous small industrial cities
or government centres. Although high on the community facilities scale,
Saint John and Campbellton are rated near the bottom of the average
group in terms of housing quality. Most of the larger cities fall in the
average category, especially on the housing list. On the community facilities list few cities seem to be 'average,' the majority being 'good' or 'poor.'
The lowest category is dominated by the Nova Scotia cities of the coal-
bearing areas where prosperity is at a low ebb as the result of many mine
closures. In both Corner Brook and Grand Falls–Windsor there exist high
quality, former company towns standing adjacent to very low quality
housing areas that began as shanty towns outside the boundaries.

The Atlantic Development Board study draws attention to the serious
shortage of housing in Atlantic Provinces cities and the rather poor quality
of housing there compared with that existing in Ontario and other parts of
Canada. New housing is built, but all too frequently without the modern
plumbing and heating facilities considered essential elsewhere. Many

Table 4.10   Indexes of housing quality and community facilities
(Source: *Urban Centres in the Atlantic Provinces*
Atlantic Development Board, pp. 70 and 78)

	Housing quality		Community facilities	
	City	Index	City	Index
Good	Oromocto	8	Oromocto	9
	Bathurst	12	Fredericton	15
	Edmundston	19	Bathurst	16
	Fredericton	20	Campbellton	18
	Charlottetown	20	*Saint John*	18
			Lancaster	23
			Charlottetown	19
			Summerside	19
Average	Moncton	22	Edmundston	20
	Halifax	22	Truro	21
	Dartmouth	24	*Halifax*	21
	New Glasgow	22	Dartmouth	26
	Stellarton	25	Corner Brook	26
	Summerside	26	St John's	27
	St John's	27	Moncton	27
	Truro	28		
	Amherst	29		
	Saint John	32		
	Lancaster	23		
	Campbellton	31		
	Grand Falls	16		
	Windsor	49		
Poor	*Sydney*	24	Amherst	31
	North Sydney	31	*Grand Falls*	30
	Glace Bay	36	Windsor	34
	New Waterford	36	*New Glasgow*	32
	Sydney Mines	43	Stellarton	34
	Corner Brook	44	*Sydney*	24
			Glace Bay	32
			North Sydney	36
			New Waterford	36
			Sydney Mines	39

people who are employed in cities built houses in small, sometimes quite distant, communities in order to evade the strict building standards of the urban areas. In this manner, the region has been solving its housing problems by sacrificing quality. A better approach, according to the Atlantic Development Board, would be land assembly under government auspices to make available serviced urban land for individual development at more modest prices than prevail in the private real estate market.

POLITICAL PATTERNS

The multiplicity of municipal jurisdictions within single urban entities has long been a problem in Atlantic Provinces cities, but it has been ameliorated in recent years by amalgamations in several principal cities. The problem is most acute in Sydney. Owing to its dispersed character, Sydney has half a dozen incorporated towns or cities, as well as a dozen unincorporated communities, making up the major urban area of slightly over 100,000 population (Figure 4.6). The cities of Halifax and Dartmouth remain separate, but in January 1969 the City of Halifax annexed approximately twenty square miles of adjacent suburban areas. Most of the built-up area now is included within the two cities. Boundary changes in both St John's and Corner Brook during the past fifteen years have included most of the urbanized areas within these cities, which greatly facilitates city management and planning. In Saint John an amalgamation of several separate jurisdictions was effected in 1967 as one of the sweeping local government changes instituted in New Brunswick. Accordingly, the trend has been towards amalgamation and, failing this, towards regional co-ordination of certain functions. An example is the recent formation of the Halifax–Dartmouth and Halifax County Regional Planning Commission.

New Brunswick has pioneered a revolutionary new approach to provincial–municipal relations that will strongly influence the cities and their future development. A Royal Commission was established in 1962 to examine the problems and recommend solutions. The recommendations of the Byrne Commission were summarized recently by Krueger:

Very briefly, the Byrne Commission made the following 'package' of proposals: that all general services to people (education, justice, health, and welfare) be financed and administered by the provincial government, and that all local services associated with property (e.g., fire protection, garbage collection, sewerage, water, parks, and community planning) be provided by local governments; that the provincial government take over all assessment and tax collection and pay equalization grants to municipalities; that county governments be abolished; and that provincial administrative commissions be established. (Krueger 1970)

Most of these recommendations were implemented in 1967. The most revolutionary changes were that services to people be the financial responsibility of the provincial government and that services to property be the financial responsibility of local government. Amalgamation of adjacent municipalities was not carried out, except in the case of Saint John. Instead, the Commission had recommended the provision of certain services, such as water, sewage disposal, fire and police protection, and

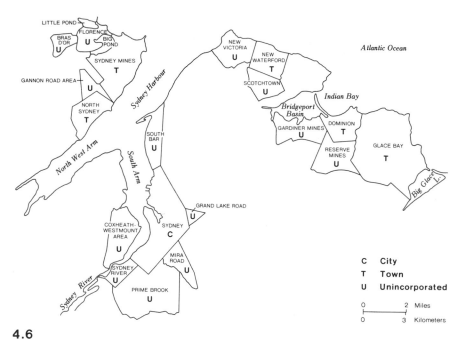

4.6

The Sydney-Glace Bay Major Urban Area 1966

community planning, on a regional basis where adjacent communities exist. Very few joint services have been established since the recommendations were adopted. According to Krueger, the most serious criticism of the program is the lack of rationalization of municipal boundaries and the lack of a regional planning framework. Nevertheless, it has introduced a rational and fair taxation system and provided all citizens with an equal opportunity to enjoy the government services to people and to property. Hopefully, the disparities between New Brunswick cities in quality of community facilities will, through a period of time, be minimized.

URBAN RENEWAL

During the past fifteen years a number of urban renewal projects have been undertaken in Atlantic Provinces cities, in recognition of the need for improvements and in view of the availability of funds. Under the National Housing Act substantial financial assistance to municipalities for such projects was available from federal and provincial governments. Most projects were designated in areas of obsolete and substandard housing and involved the erection of new residential structures, along with some offices and stores or government buildings. Many of the dwelling units were built as subsidized rental accommodation for low-income families, with the assistance of the Central Mortgage and Housing Corporation, a federal government agency. Toward the end of the 1960s renewal activity tapered off because the federal government cut back financial support pending a complete review of the program. One of the problems was that the program, as it operated through the 1950s and 1960s, usually resulted in the complete demolition of all structures on a site and the erection of antiseptic, institutional-type apartments or row houses that did not harmonize with the older buildings around them. There was insufficient emphasis upon rehabilitation and selective renewal in an attempt to retain the character of a neighbourhood. The disadvantages of the program were less serious in Atlantic Provinces cities than in many newer cities to the west. In fact, the old cities of Halifax, Saint John, and St John's had areas of such dilapidated housing that complete demolition was the only alternative. Also, the development of well-planned, renewal complexes, including residential, commercial, and governmental functions, proved successful in several cities of the Atlantic Provinces.

The 1957 urban renewal study in Saint John reported that nearly one-third of the dwelling units in the city were in such a state of deterioration as to warrant immediate renewal. Courtenay Place, begun in 1961, was the first renewal project which now houses 190 families, a recreation facility, and a number of commercial functions on a 60-acre site. The North End project, including some 200 acres around the head of the harbour, followed in the mid-1960s. Still under development, it will include a greatly improved circulation system, favourable sites for offices and commercial establishments, and new areas for apartments. An important benefit will be the opening up of an attractive waterfront route to downtown from West Saint John.

Halifax has been very active in the field of urban renewal since the late 1950s. In fact, the first non-residential redevelopment in Canada was begun in Halifax at that time (Pickett 1958). A 12-acre site near the City Hall was cleared with the intention of reuse for commercial purposes. After lying fallow for a number of years the site was finally developed as Scotia Square. Residential renewal was advanced far more rapidly, and by 1965 there were approximately 700 new dwelling units of subsidized rental housing. While Halifax contained less than one-half of one per cent of the total number of housing units in Canada, it contained 7.3 per cent of the nation's subsidized housing (*Housing Report No. 2*, Halifax, 1965). One of the largest residential projects is Uniacke Square situated north of the central business district which, when completed in its entirety, will make available about 1100 subsidized dwelling units. Over a period of several years the Black community of Africville at the north end of the Halifax peninsula was moved into new housing on a renewal site. The old shacks in this former ghetto were torn down to prepare the land for non-residential uses.

Development of the cleared area north of the central business district, which was expanded to 17 acres by 1965, was beset with frustration. The several goals of the project were to remove the blighted residential structures that housed 270 families, to redevelop the street system, with provision for a thoroughfare of moderate grade from the harbour up to the foot of the Citadel, and to expand the commercial zone with the hope of rejuvenating the central business district. The first private development proposal accepted did not reach fruition, and further studies of the central business district were undertaken, culminating in the acceptance in 1966 of the Scotia Square proposal (*City of Halifax Central Business District, Phase 2 – Redevelopment Planning and Implementation*, 1966; and *Report of City of Halifax Joint Staff Review Committee*, 1966). The forty million dollar project consists of several buildings linked at lower levels including: five apartment buildings, two office towers, a hotel, a theatre, a wholesale trade mart, a two-level enclosed shopping mall, and underground parking for 1600 cars. The mall has two department stores, a supermarket, and 70 specialty shops. While incomplete as yet, the first phase was opened in 1969 and already has proven a great success. This project has generated an excitement in Halifax that is comparable in many ways with the tremendous boost that Montreal experienced with the building of Place Ville Marie at the beginning of the 1960s. Another renewal project is underway nearby on the waterfront, where a new Court House and plaza are being constructed. There is an attempt to link these two projects with a landscaped mall along Granville Street, where old buildings dating from 1855–1895 would be renovated for retail, office, and apartment uses.

The most spectacular renewal project in St John's was the twenty million dollar redevelopment of the harbour by the federal Department of Public Works. Begun in 1958 and completed in 1967, it involved the building of new wharves around a large part of the harbour perimeter. The north side of the harbour bordering the central business district was the zone of major redevelopment. A marginal wharf replaced the former H.M.C. Dockyard and about forty small, obsolete finger wharves. It was backed by a broad access roadway along the waterfront where none existed before. This changed the face of downtown St John's, opening up the harbour to the public, and providing an alternate route for traffic parallel to Water Street. In fact, it made possible the closing of a portion of Water Street for a shopping mall.

Residential renewal in St John's dates from the mid-1950s and considerable emphasis has been placed on the provision of subsidized housing. Even by 1960 there were nearly 600 such dwelling units completed (*Urban Renewal Study* 1961). As occurred in Halifax, a large area on the edge of downtown was cleared for renewal but remained vacant for many years. The new City Hall, along with residential buildings, now occupies this site. Rehabilitation and renewal of fringe communities form another aspect of the program in St John's. The Blackhead community, with a population of about 1400, is an interesting example. It is situated on a high ridge south of the harbour, just outside the former city boundary, though only a mile from downtown. It is reached by an extremely steep road beginning at the head of the harbour. The community originated during the 1930s when migrants moving to St John's from the outports in search of employment were unable to find accommodation. They were permitted by the provincial government to settle on Crown Land on the upper slopes of the South Side Hills, where they built homes with second-hand lumber. There were minimal services and housing quality, which was mediocre to begin with, deteriorated rapidly. The feasibility of providing services for the area was confirmed in 1965 and the project went ahead, involving rehabilitation and limited renewal of structures.

THE FUTURE

As places to live, the Atlantic Provinces cities are pleasant and less hectic than larger cities elsewhere in Canada. There is every likelihood that these conditions will continue to prevail, owing to the slower pace of growth. Because the Atlantic Provinces economy has consistently lagged behind the other regional economies in Canada, however, standards of living in these cities are lower than in most other Canadian cities. A corollary of this is that unemployment and housing problems are especially difficult to solve.

Great efforts have been made by the Atlantic region provincial governments and by the federal government to institute programs with the aim of fostering economic growth towards a closing of the standard of living gap between the Atlantic region and the rest of Canada. Approaches to the problem have been varied. The Atlantic Development Board has placed heavy emphasis upon the development of infrastructure and has made available substantial grants for this purpose. The Department of Regional Economic Expansion has designated the whole of the Atlantic Provinces region as eligible for industrial incentives. Other organizations have con-centrated on particular regions, for example, the Fund for Regional Economic Development has concentrated on Prince Edward Island and northeastern New Brunswick, while the Cape Breton Development Cor-poration has been concerned with the island portion of Nova Scotia. The provinces themselves have established industrial development agencies, such as Industrial Estates Limited in Nova Scotia and the New Brunswick Development Corporation. A great many of these governmental activities have been 'scatter gun' approaches without master plans for concentration of development in specified centres. Indeed, the industrial incentives pro-gram of the Department of Regional Economic Expansion formerly de-clared the urban areas of Halifax–Dartmouth, Saint John, and Fredericton as ineligible for grants because they did not have unemployment levels high enough to meet the criteria applied. In order to counteract the frag-mentation problem of the region, greater co-ordination and integration of development are needed.

The establishment of growth centres has been suggested as one viable strategy for development in the region (*A Development Program for the Atlantic Provinces* 1965, and *Urban Centres in the Atlantic Provinces* 1969). The centres of sub-regions which have the best advantage for growth and concentration of activities have been identified. The Atlantic Provinces Economic Council suggests that Saint John, Moncton, Bathurst–Belledune, Charlottetown, Halifax, Truro, Sydney, Corner Brook and St John's are most appropriate as key development centres. Going one step beyond this recognition of sub-regional centres, it seems essential that there be one major metropolis within the region. A strong regional capital is required as a focal point in a better articulated urban system.

Halifax is the logical choice because it is already well on the way towards achieving this role. It is far enough away from Montreal and Toronto to benefit from the advantage of isolation in terms of carving out a regional market. At the same time, Halifax is centrally located within the whole, attenuated Atlantic Provinces region, even though it is distinctly off-centre in terms of the Maritimes alone. A conscious effort by the federal and Nova Scotia governments to concentrate development in Hali-fax whenever possible could initiate growth on the scale required. It has been observed that once a city reaches a size of approximately half a million population its growth becomes self-sustaining, even in the face of economic adversity. The emphasis need not be on industrial development, but rather on tertiary activities, including finance, trade, services, and government. The site of metropolitan Halifax offers ample space for urban expansion, despite localized occurrences of unfavourable terrain in the vicinity (*Report on Regional Development District of Halifax, Dartmouth and Halifax County* 1970). Also in its favour is its excellent harbour, offering deep enough water to accommodate the new generation of large ships and promising to become one of North America's leading container ports.

Other growth centres would function as sub-regional central places, as they do at present. A greatly improved transportation network between centres would enhance interaction and foster growth of the larger cities. In New Brunswick the urban functions are so divided already that there seems little likelihood of one centre becoming strongly dominant within the province. Saint John will remain the largest for some time to come, but its low rate of growth may continue. It enjoys favourable prospects as a port, especially in the container trade, and has a strong industrial base on which to build. Moncton has an assured role as an important central place and transportation centre, and may experience a faster growth rate than Saint John. Fredericton can look forward to steady growth through its governmental function, but Charlottetown probably will grow very slowly, owing to its restricted island hinterland. Although strong efforts are being made to diversify the industrial economy of Sydney, the prospect is that this urban area will experience the lowest population growth of all the major cities in the near future. St John's and Corner Brook have few rivals to challenge their effective division of the island of Newfoundland into two tributary areas for higher-order central place functions. Corner Brook possesses locational advantages relative to mainland sources of supply and St John's has the weight of tradition, governmental control, economic power, and sheer size on its side.

ACKNOWLEDGMENT

The research grant support of the Canada Council which enabled the author to visit several Atlantic Provinces port cities is gratefully acknowledged.

Bibliography

Chapter 1

Alcock, F.J., 1928   Rivers of Gaspé, *Bull. Geol. Soc. Amer.*, 39: 403–420

— 1935   Geology of the Chaleur Bay Region, *Geol. Surv. Canada, Memoir* 183

Atlantic Development Board, 1968   *Forestry in the Atlantic Provinces.* Background Study No. 1 (Ottawa)

Bartlett, G.A. and L. Smith, 1971   Mesozoic and Cenozoic History of the Grand Banks of Newfoundland. *Can. J. Earth Sci.*, 8: 65–84

Bird, J.B., 1964   *The Uplands of the Canadian Maritime Provinces*, unpublished MS

Black, W.A., 1959*a*, 1959*b*, 1960, 1961, 1963   Gulf of St. Lawrence Ice Survey, *Geographical Branch Papers* 19 (Winter 1958), 23 (Winter 1959), 25 (Winter 1960), 32 (Winter 1961), 36 (Winter 1962) (Ottawa)

Brookes, I.A., 1964   *The Upland Surfaces of Western Newfoundland*, unpub. M.Sc. thesis, Department of Geography, McGill University

— 1970   New evidence for an independent Wisconsin-age ice cap over Newfoundland. *Can. J. Earth Sci.*, 7: 1374–1382

Canada, Department of Transport, 1968*a*   Climatic Normals. i. Temperature. Meteorological Branch (Toronto)

— 1968*b*   Climatic Normals. ii. Precipitation. Meteorological Branch, Toronto, 110 p.

Clark, A.H., 1968   *Acadia: The Geography of Early Nova Scotia to 1760* (Madison, Wis.)

Dawson, J.W., 1891   *The Geology of Nova Scotia, New Brunswick, and Prince Edward Island, or Acadian Geology*, 4th ed., with supplementary note, and supplement to 2nd ed. of 1878 (Edinburgh)

Douglas, R.J.W., 1969   Geological Map of Canada. *Geol. Surv. Canada,* Map 1250A

Dunn, C.W., 1953   *The Highland Settler* (Toronto)

Erskine, D.S., 1961   Plants of Prince Edward Island. *Canada Dept. Agric. Research Branch, Publication* 1088

— 1968   The Atlantic Region, *in* John Warkentin (ed.), *Canada: A Geographical Interpretation* (Toronto)

Fernow, B.E., 1912   Forest Conditions of Nova Scotia, *Canada, Commission of Conservation* (Ottawa)

Forward, C.N., 1954   Ice distribution in the Gulf of St. Lawrence during the break-up season. *Geog. Bull.*, 6: 45–84

Gadd, N.R., 1970   Quaternary geology, southwest New Brunswick, *in* Report of Activities, *Geol. Surv. Canada Paper*, 70–1A

Goldthwait, J.W., 1924   Physiography of Nova Scotia. *Geol. Surv. Canada, Memoir* 140

Hachey, H.B., 1961   Oceanography and Canadian Atlantic Waters. *Canada, Fish. Res. Bd., Bull.* 134

Howden, H.F., J.E.H. Martin, E.L. Bousfield, and D.E. McAllister, 1970 Fauna of Sable Island and its zoogeographic affinities. *Nat. Mus. Canada, Publications in Zoology*, No. 4

King, L.H., 1969 Submarine end moraines and associated deposits on the Scotian Shelf, *Bull. Geol. Soc. Amer.*, 89: 83–96

King, L.H. and B. MacLean, 1970 Origin of the outer part of the Laurentian Channel. *Can. J. Earth Sci.*, 7: 1470–1484

King, L.H., B. MacLean, G.A. Bartlett, J.A. Jeletzky, and W.S. Hopkins, Jr., 1970 Cretaceous strata on the Scotian Shelf. *Can. J. Earth Sci.*, 7: 145–155

Leim, A.H. and W.B. Scott, 1966 Fishes of the Atlantic Coast of Canada. *Fish. Res. Bd., Bull.*, 155–465

Livingstone, D.A., 1968 Some interstadial and postglacial pollen diagrams from eastern Canada. *Ecol. Monog.*, 38: 87–125

Loucks, O.L., 1962 A Forest Classification for the Maritime Provinces, *Proc. Nova Scotia Inst. Sci.*, 25 (2), 86–167

Mollard, J.D. and L.C. Munn, 1956 *Report on Classes on Land in Newfoundland*. Appendix 1, *in* Report of the Royal Commission on Agriculture. St. John's, Newfoundland

New Brunswick Soil Survey. Listed by area, report number, year. Woodstock, 2, 1944 (published as Can. Dept. Agric. Pub. 757, Tech. Bull. 48); Southeastern New Brunswick, 3, 1949; Southwestern New Brunswick, 4, 1953; Andover-Plaster Rock, 5, 1964

Nova Scotia Soil Survey. Reports listed here by county, report number and year. Cumberland, 2, 1945; Colchester, 3, 1948; Pictou, 4, 1950; Hants, 5, 1954; Antigonish, 6, 1954; Lunenburg, 7, 1958; Queens, 8, 1959; Yarmouth, 9, 1960; Shelburne, 10, 1961; Digby, 11, 1962; Cape Breton Island, 12, 1963; Halifax, 13, 1963; Guysborough, 14, 1964; Kings, 15, 1965; Annapolis, 16, 1969

Ogden, J.G. III, 1965 Pleistocene pollen records from eastern North America. *Bot. Review*, 31; 481–504

Poole, W.H., B.V. Sandford, H. Williams, and D.G. Kelley, 1970 Geology of Southeastern Canada, *in* R.J.W. Douglas (ed.), *Geology and Economic Minerals of Canada*, Econ. Geol. Rept. No. 1, 5th Ed. (Ottawa)

Prest, V.K., 1969 Retreat of Wisconsin and Recent Ice in North America, *Geol. Surv. Canada*, Map 1257A

— 1970 Quaternary Geology of Canada, *in* R.J.W. Douglas (ed.), *Geology and Economic Minerals of Canada*, Econ. Geol. Dept., No. 1, 5th Ed. (Ottawa)

Prest, V.K., D.R. Grant, and V.N. Rampton, 1968 Glacial Map of Canada, *Geol. Surv. Canada*, Map 1253A

— 1969 Retreat of the last ice sheet from the Maritime Provinces – Gulf of St. Lawrence region. *Geol. Surv. Canada*, Paper 69–33

Roland, A.E. and E.C. Smith, 1969 The Flora of Nova Scotia. Part II. The Dicotyledons. *Proc. Nova Scotia Inst. Sci.*, 26 (4): 277–743

Rowe, J.S., 1959   Forest Regions of Canada. *Canada, Dept. Northern Affairs and National Resources, Forestry Branch*, Bull. 123

Shepard, F.P., 1963   Thirty-five thousand years of sea level, *in Essays in Marine Geology in Honour of K.O. Emery* (Los Angeles)

Stockwell, C.H. (chairman), 1969   Tectonic Map of Canada, *Geol. Surv. Canada*, Map 1251A

Templeman, W., 1966   Marine Resources of Newfoundland, *Fish. Res. Bd., Bull.*, 154

Twenhofel, W.H., 1912   The Physiography of Newfoundland. *Am. J. Sci.*, 4th series, 33 (193): 1–25

Twenhofel, W.H. and P. MacClintock, 1940   Surface of Newfoundland, *Bull. Geol. Soc. Am.*, 51: 1665–1728

Whiteside, G.B., 1950   Soil Survey of Prince Edward Island. Charlottetown, *Experimental Farms Service*

Whitmore, F.C. Jr., K.O. Emery, H.B.S. Cooke, and D.J.P. Swift, 1967   Elephant teeth from the Atlantic continental shelf. *Science*, 156: 1477–1480

Chapter 2

Brown, R., 1968   *Surname, denominational affiliation and occupational structure, Joe Batt's Arm, Newfoundland*, unpublished maps and report, Department of Geography, Memorial University

Clark, A.H., 1959   *Three Centuries and the Island: a historical geography of settlement and agriculture in Prince Edward Island* (Toronto)

— 1968   *Acadia: the geography of early Nova Scotia to 1760* (Wisconsin)

Courtney, D.S., 1972   *Newfoundland Resettlement: a case study in spatial re-organisation and growth centre strategy*, unpublished report, Department of Geography, Memorial University

Delaney, R.E., *Kinship and Settlement Morphology in southwestern Newfoundland: a study of selected Highland Scottish settlements.* MA thesis, Memorial University (in preparation)

Delaney, R.E. and A.G. Macpherson, 1971   *The role of the Scot in the historical geography of Newfoundland*, paper presented to the Conference of Scottish Studies, St John's, Newfoundland

Department of Energy, Mines and Resources, Canada, 1968   *Socio-economic interview statistics, Pictou County, Nova Scotia* (in co-operation with A.R.D.A.), Truro, N.S. (unpub.)

Department of Regional Economic Expansion, Canada, 1970   *Population characteristics: Unincorporated Communities, Newfoundland and Labrador* (Ottawa)

Dominion Bureau of Statistics, Canada, *Census of Canada, 1961* (Ottawa)

— *Census of Canada, 1966* (Ottawa)

Gentilcore, R.L., 1956   The agricultural background of settlement in eastern Nova Scotia, *Annals Assoc. Amer. Geogr.*, 46: 378–404

Government of Newfoundland and Labrador, *Federal-Provincial Resettlement Program, Newfoundland. Statistics, first five-year period: April 1, 1965–March 31, 1970* (St John's)

Iverson, N. and D.R. Matthews, 1968 *Communities in Decline: an examination of household resettlement in Newfoundland*, Newfoundland Social and Economic Studies, 6, Institute of Social and Economic Research, Memorial University

Jackson, C.I. and J.W. Maxwell, 1971 *Landowners and Land Use in the Tantramar Area, New Brunswick*, Canada Land Inventory Report, 9, Department of Regional Economic Expansion, and Geographical Paper, 47, Department of Energy, Mines and Resources (Ottawa)

Lycan, R., 1969 Interprovincial Migration in Canada: the role of spatial and economic factors, *Can. Geogr.*, 13: 237–254

Mannion, J.J., 1971 *Irish Imprints on the Landscape of Eastern Canada in the Nineteenth Century: a study in cultural transfer and adaptation*, unpublished PHD thesis, Department of Geography, University of Toronto

Noel, S.J.R., 1971 *Politics in Newfoundland* (Toronto)

Raymond, C.W. and J.A. Rayburn, 1963 *Land abandonment in Prince Edward Island*, Geographical Bulletin, 19, Geographical Branch, Department of Mines and Technical Surveys, Canada (Ottawa)

Skolnik, M.L. (ed.), 1968 *Viewpoints on Communities in Crisis*, Newfoundland Social and Economic Papers, 1, Institute of Social and Economic Research, Memorial University

Statistics Canada, *Census of Canada, 1971; Preliminary Bulletin*, 3 (Ottawa)

Thorburn, H.G., 1961 *Politics in New Brunswick* (Toronto)

Wadel, C., 1969 *Marginal Adaptations and Modernization in Newfoundland: a study of strategies and implications in the resettlement and redevelopment of outport fishing communities*, Newfoundland Social and Economic Studies, 7, Institute of Social and Economic Research, Memorial University

Chapter 3

Atlantic Development Board, 1968 Forestry in the Atlantic Provinces, *Background Study* 1 (Ottawa), 270

— 1969 Fisheries in the Atlantic Provinces, *Background Study* 3 (Ottawa)

— 1969 Mineral resources in the Atlantic Provinces, *Background Study* 4 (Ottawa), 94

Atlantic Provinces Economic Council, 1966 Agriculture and the Atlantic economy, *Pamphlet* 10 (July) (Fredericton), 44

— 1968 Atlantic Provinces Fishery. *Pamphlet* 12 (June) (Fredericton), 55

Atlantic Tidal Power Programming Board, 1969 *Feasibility of tidal power development in the Bay of Fundy* (Ottawa), 31

Black, W.A., 1972 *The View from Water Street*, Social Science Series, No. 2, Inland Waters Directorate, Department of the Environment (Ottawa), 20

Boni, Watkins, Jason & Co., Inc., 1969 *Economic Outlook for the Province of Newfoundland* (St John's), 115

Booth, J.F., G.C. Retson, and V.A. Heighton, 1970   *The Agriculture of the Atlantic Provinces*, Department of Regional Economic Expansion (Ottawa)

Churchill Falls (Labrador) Corp. Ltd. (n.d.)   *Power from Labrador – the Churchill Falls Development* (St John's), 50

(Davis, Hon. J.), 1969   Outlines government proposals to re-organize fisheries, *Fisheries of Canada* (Ottawa: Department of Fisheries and Forestry), 22, 6 (Dec.): 3–5

Department of Fisheries and Forestry, 1969   Pollution and the fisheries act. *Fisheries of Canada* (Ottawa: Department of Fisheries and Forestry), 22, 1 (July): 3–5

— 1971   Exclusive fishing zones proclaimed by Canada, *Fisheries of Canada* (Ottawa) 23, 4 (March–April): 10–12

Energy Development Sector, 1969   *Electric power in Canada* (Ottawa: Department of Energy, Mines and Resources), 88

Government of Newfoundland, 1964   *National fisheries development: the presentation to federal-provincial conference on fisheries development* (St John's): 44

Hilchey, John D., 1970   Soil capability analysis for agriculture in Nova Scotia, *The Canada Land Inventory*, 8 (Ottawa: Department of Regional Economic Expansion), 66

Lafleur, P., 1970   *Iron ore*, 22 (Ottawa: Department of Energy, Mines and Resources), 17

Maxwell, J.W., N.J. Sagar, and D.K. Redpath, 1967   *Atlantic Provinces resources and economic activity* (Ottawa: Department of Energy, Mines and Resources) (Map)

Mitchell, C.L., 1969   Atlantic coast herring fishery, *Fisheries of Canada* (Ottawa: Department of Fisheries and Forestry), 2, 6 (Dec.): 13–18

Nowlan, D.M., 1962   The demand for energy in the Atlantic Provinces, 1950–1980, *Information Bulletin*, 57 (Ottawa: Department of Mines and Technical Surveys), 114

Report of the Royal Commission on the Economic State and Prospects of Newfoundland and Labrador, 1967 (St John's), 499

Chapter 4

A Development Program For The Atlantic Provinces, 1965   Pamphlet No. 8, Atlantic Provinces Economics Council (Halifax)

Air Passenger Origin and Destination, 1969   Domestic Report, Dominion Bureau of Statistics (Ottawa)

A Service-based Community: The Example of Fredericton-Oromocto, Pamphlet No. 14, 1969   Atlantic Provinces Economics Council (Halifax)

City of Halifax Central Business District, Phase #2 – Redevelopment Planning and Implementation, 1966   Canadian Urban Economics Limited (Toronto)

City of Halifax Industrial Park Site Analysis, 1969   Canadian National (Moncton)

Clark, A.H., 1959   *Three Centuries and The Island* (Toronto)

Coblentz, H.S., 1963   *Halifax Region Housing Survey* (Halifax)

Forward, C.N., 1967   Recent Changes in the Form and Function of the Port of St. John's, Newfoundland, *Canadian Geographer*, XI, 2: 101–116

— 1967   Port Functions and Shipping Trade of Halifax, Nova Scotia, *in Geographical Perspectives: Some Northwest Viewpoints*, B.C. Geographical Series, No. 8, Occasional Papers in Geography, Tantalus Research (Vancouver)

— 1969   A Comparison of Waterfront Land Use in Four Canadian Ports: St. John's, Saint John, Halifax and Victoria, *Economic Geography*, 45, 2: 155–169

Housing Report 2, 1965   Development Plan Series (Halifax)

Industrial Park Needs of the Metropolitan Area of Saint John, 1966   Stevenson and Kellogg Ltd. (Halifax)

Kerr, Donald, 1965   Some Aspects of the Geography of Finance in Canada, *Canadian Geographer*, IX, 4: 175–192

Krueger, R.R., 1970   The Provincial-Municipal Government Revolution in New Brunswick, *Canadian Public Administration*, 13, 1: 51–99

Maxwell, J.W., 1965   The Functional Structure of Canadian Cities: A Classification of Cities, *Geographical Bulletin*, 7, 2: 79–104

Pearson, R.E., 1969   *Atlas of St. John's, Newfoundland*, Department of Geography, Memorial University of Newfoundland (St. John's)

Pickett, S.H., 1968   *Urban Renewal*, Pamphlet No. 1, Community Planning Association of Canada (Ottawa)

Prowse, D.W., 1895   *A History of Newfoundland*, Macmillan and Co. (London)

Raddall, T.H., 1965   *Halifax, Warden of the North*, Doubleday (New York)

Raymond, W.O., 1950   *The River St. John*, 2nd ed., Tribune Press (Sackville)

Report of City of Halifax Joint Staff Review Committee, 1966   (Halifax)

Report on Regional Development, District of Halifax, Dartmouth and Halifax County, 1969   Murray V. Jones and Associates Limited (Toronto)

Sinclair, Alasdair M., 1961   *The Economic Base of the Halifax Metropolitan Area*, Institute of Public Affairs, Dalhousie University (Halifax)

Somerville, M.M., 1968   Urban Renewal, *Community Planning Review*, 18, 4: 5–8

Stone, Leroy O., 1967   *Urban Development in Canada*, 1961 Census Monograph, Dominion Bureau of Statistics (Ottawa)

Urban Centres in the Atlantic Provinces, 1969   Background Study No. 7, Atlantic Development Board (Ottawa)

Urban Renewal Study, City of St. John's, Newfoundland, 1961   Project Planning Associates Limited (Toronto)

Winks, R.W., 1971   *The Blacks in Canada*, McGill-Queen's University Press and Yale University Press (Montreal)